Trigonometry Essentials Practice Workbook with Answers

Master Basic Trig Skills

Improve Your Math Fluency Series

Chris McMullen, Ph.D.

Trigonometry Essentials Practice Workbook with Answers: Master Basic Trig Skills

Improve Your Math Fluency Series

Copyright © 2012, 2017 Chris McMullen, Ph.D.

CreateSpace

Professional & Technical / Science / Mathematics / Trigonometry
Professional & Technical / Education / Specific Skills / Mathematics / Trigonometry

ISBN: 1477497781

EAN-13: 978-1477497784

Contents

Making the Most of this Workbook

- Mathematics is a language. You can't hold a decent conversation in any language if you have a limited vocabulary or if you are not fluent. In order to become successful in mathematics, you need to practice until you have mastered the fundamentals and developed fluency in the subject. This *Trigonometry Essentials Practice Workbook with Answers: Master Basic Trig Skills* will help you improve the fluency with which you apply fundamental trig techniques. Every problem can be answered without a calculator, which is very helpful for students who aren't allowed to use a calculator. This is the case in some trig and physics courses, as well as some standardized exams (like the MCAT).

- This workbook is conveniently divided into 11 chapters so that you can focus on one basic skill at a time. The first two chapters provide practice converting between degrees and radians. Chapters 3-4 are devoted toward relating the basic trig functions to right triangles; Chapter 4 focuses on the 30°-60°-90° and 45°-45°-90° triangles. Students master the basic trig functions in Chapters 5-7, and their inverse functions in Chapter 8. Chapter 9 is devoted to the law of sines and law of cosines, and includes several obtuse and acute triangles for practice. Trig identities and their application are the subject of Chapter 10, and Chapter 11 focuses on how to solve equations that feature trig functions.

- Each chapter begins with concise instructions describing how to perform a basic trig skill – such as how to determine the reference angle. These instructions are followed by a few examples. Use these examples as a guide until you become fluent in the technique.

- After you complete a page, check your answers with the answer key in the back of the book. Practice makes permanent, but not necessarily perfect: If you practice making mistakes, you will learn your mistakes. Check your answers and learn from your mistakes such that you practice solving the problems correctly. This way your practice will make perfect.

- Math can be fun. Make a game of your practice by recording your times and trying to improve on your times, and recording your scores and trying to improve on your scores. Doing this will help you see how much you are improving, and this sign of improvement can give you the confidence to succeed in math, which can help you learn to enjoy this subject more.

Chapter 1: Converting Degrees to Radians

There are two common methods for measuring angles. One method is to divide a circle up into 360 slices, where each slice equals one degree (°). In the degree measure, a full circle corresponds to 360°, a right angle is 90°, an equilateral triangle has 60° angles, and so on. Degrees are very common in science and engineering, since a protractor is typically ruled in degrees.

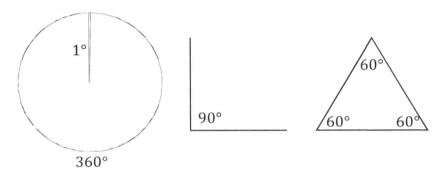

A second method for measuring angles is to work with radians instead of degrees. Radians are defined such that 2π radians correspond to a full circle. The unit radian is often abbreviated rad. In terms of radians, a right angle is $\pi/2$ rad, an equilateral triangle has angles of $\pi/3$ rad, and so on. Radians are very common in math courses since many geometric formulas involve π.

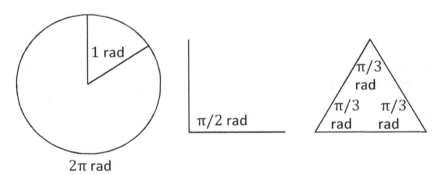

Since both units for angular measure – degrees and radians – are very common, it's useful to be able to convert degrees to radians or vice-versa.

The main idea behind the conversion is geometric. For example, 60° is one-sixth of a circle, since 60° equals 360°/6. Therefore, the same angle in radians is $\pi/3$ rad, since one-sixth of 2π is $2\pi/6$, which reduces to $\pi/3$. Similarly, 90° is one-fourth of a circle, since 360° divided by 4 equals 90°. This equates to $\pi/2$ rad, since 2π divided by 4 reduces to $\pi/2$.

The conversion factor needed to convert degrees to radians or vice-versa is:

$$180° = \pi \text{ radians}$$

The reason for this is that a full circle is 360° or 2π radians. Therefore, 360° is equivalent to 2π radians. If we divide both quantities by 2, we find that 180° equates to π radians.

Dividing both sides by π, you can see that 1 radian equates to approximately 57.3 degrees. However, it's usually more convenient to remember that 180° = π radians than to memorize the number 57.3.

In this chapter, we will practice converting degrees into radians. The way to do this is to multiply by π radians and divide by 180°, as illustrated in the following examples.

Take some time to understand these concepts, and study the examples. Once you understand the following examples, you are ready to practice the technique yourself. You may need to refer to the examples frequently as you begin, but should try to solve the exercises all by yourself once you get the hang of it. Be sure to check the answers at the back of the book to ensure that you are solving the problems correctly.

Instructions: Convert the given angle from degrees to radians.

Procedure: Multiply the given angle by π and divide by 180. If the result is a fraction, see if the fraction is reducible. That is, if the numerator and denominator are both evenly divisible by an integer greater than 1, then the fraction is reducible. Reduce a fraction by dividing both the numerator and denominator by the greatest common factor (as in the examples that follow).

Example 1: 120°

$$120° \times \frac{\pi \text{ rad}}{180°} = \frac{120}{180}\pi \text{ rad} = \frac{2\pi}{3} \text{ rad}$$

Both the numerator (120) and denominator (180) are divisible by 60: 120/60 = 2, 180/60 = 3. The answer is $2\pi/3$ rad.

Example 2: 30°

$$30° \times \frac{\pi \text{ rad}}{180°} = \frac{30}{180}\pi \text{ rad} = \frac{\pi}{6} \text{ rad}$$

Both the numerator (30) and denominator (180) are divisible by 30: 30/30 = 1, 180/30 = 6. The answer is $\pi/6$ rad.

Example 3: 720°

$$720° \times \frac{\pi \text{ rad}}{180°} = \frac{720}{180}\pi \text{ rad} = 4\pi \text{ rad}$$

Both the numerator (720) and denominator (180) are divisible by 180: 720/180 = 4. The answer is 4π rad.

Instructions: Convert the given angle from degrees to radians. Check your answers in the back of the book.

(1) 36°

(2) 18°

(3) 780°

(4) 117°

(5) 45°

(6) 12°

(7) 96°

(8) 270°

(9) 120°

(10) 78°

(11) 48°

(12) 150°

Instructions: Convert the given angle from degrees to radians. Check your answers in the back of the book.

(1) 210°

(2) 45°

(3) 24°

(4) 3°

(5) 10°

(6) 135°

(7) 96°

(8) 315°

(9) 270°

(10) 42°

(11) 87°

(12) 69°

Instructions: Convert the given angle from degrees to radians. Check your answers in the back of the book.

(1) 153°

(2) 48°

(3) 57°

(4) 63°

(5) 96°

(6) 552°

(7) 606°

(8) 132°

(9) 225°

(10) 300°

(11) 129°

(12) 60°

Instructions: Convert the given angle from degrees to radians. Check your answers in the back of the book.

(1) 348°

(2) 117°

(3) 105°

(4) 816°

(5) 90°

(6) 57°

(7) 27°

(8) 69°

(9) 525°

(10) 276°

(11) 63°

(12) 636°

Instructions: Convert the given angle from degrees to radians. Check your answers in the back of the book.

(1) 138°

(2) 39°

(3) 687°

(4) 273°

(5) 51°

(6) 72°

(7) 66°

(8) 288°

(9) 30°

(10) 120°

(11) 99°

(12) 144°

Instructions: Convert the given angle from degrees to radians. Check your answers in the back of the book.

(1) 510°

(2) 453°

(3) 303°

(4) 159°

(5) 507°

(6) 141°

(7) 99°

(8) 114°

(9) 102°

(10) 552°

(11) 540°

(12) 138°

Instructions: Convert the given angle from degrees to radians. Check your answers in the back of the book.

(1) 444°

(2) 813°

(3) 54°

(4) 117°

(5) 63°

(6) 435°

(7) 27°

(8) 93°

(9) 732°

(10) 51°

(11) 126°

(12) 72°

Instructions: Convert the given angle from degrees to radians. Check your answers in the back of the book.

(1) 75°

(2) 177°

(3) 708°

(4) 168°

(5) 9°

(6) 819°

(7) 93°

(8) 108°

(9) 180°

(10) 39°

(11) 366°

(12) 3°

Instructions: Convert the given angle from degrees to radians. Check your answers in the back of the book.

(1) 99°   (2) 96°

(3) 336°   (4) 324°

(5) 114°   (6) 297°

(7) 156°   (8) 360°

(9) 300°   (10) 147°

(11) 90°   (12) 60°

Instructions: Convert the given angle from degrees to radians. Check your answers in the back of the book.

(1) 162°   (2) 81°

(3) 63°   (4) 51°

(5) 66°   (6) 45°

(7) 75°   (8) 105°

(9) 78°   (10) 87°

(11) 54°   (12) 57°

Instructions: Convert the given angle from degrees to radians. Check your answers in the back of the book.

(1) 111°   (2) 159°

(3) 87°   (4) 231°

(5) 84°   (6) 9°

(7) 336°   (8) 30°

(9) 333°   (10) 171°

(11) 48°   (12) 6°

Instructions: Convert the given angle from degrees to radians. Check your answers in the back of the book.

(1) 108° (2) 153°

(3) 42° (4) 126°

(5) 306° (6) 117°

(7) 90° (8) 300°

(9) 219° (10) 93°

(11) 63° (12) 72°

Chapter 2: Converting Radians to Degrees

This chapter is just like the previous chapter, except that this time we will practice converting radians into degrees. The way to do this is to multiply by 180° and divide by π radians, as illustrated in the following examples.

Take some time to understand the concepts, and study the examples. Once you understand the following examples, you are ready to practice the technique yourself. You may need to refer to the examples frequently as you begin, but should try to solve the exercises all by yourself once you get the hang of it. Be sure to check the answers at the back of the book to ensure that you are solving the problems correctly.

Instructions: Convert the given angle from radians to degrees.

Procedure: Multiply the given angle by 180 and divide by π. Each answer in this chapter will be an integer in degrees; none of the answers in this chapter is a fraction. However, the given angle in radians may be a fraction; multiply the fraction by the specified conversion factor, as illustrated in the examples below.

Note: In this chapter, the π's will cancel.

Example 1: π/2 rad

$$\frac{\pi}{2} \text{ rad} \times \frac{180°}{\pi \text{ rad}} = \frac{180°}{2} = 90°$$

The answer is 90°.

Example 2: 3π/4 rad

$$\frac{3\pi}{4} \text{ rad} \times \frac{180°}{\pi \text{ rad}} = \frac{3 \times 180°}{4} = 135°$$

The answer is 135°.

Example 3: 5π/3 rad

$$\frac{5\pi}{3} \text{ rad} \times \frac{180°}{\pi \text{ rad}} = \frac{5 \times 180°}{3} = 300°$$

The answer is 300°.

Example 4: 3π rad

$$3\pi \text{ rad} \times \frac{180°}{\pi \text{ rad}} = 3 \times 180° = 540°$$

The answer is 540°.

Instructions: Convert the given angle from radians to degrees. Check your answers in the back of the book.

(1) $9\pi/20$ rad

(2) $\pi/20$ rad

(3) $17\pi/30$ rad

(4) $7\pi/20$ rad

(5) $29\pi/30$ rad

(6) $\pi/12$ rad

(7) $13\pi/15$ rad

(8) $47\pi/20$ rad

(9) π rad

(10) $\pi/4$ rad

(11) $139\pi/30$ rad

(12) $\pi/2$ rad

Instructions: Convert the given angle from radians to degrees. Check your answers in the back of the book.

(1) $217\pi/60$ rad

(2) $5\pi/12$ rad

(3) $17\pi/20$ rad

(4) $19\pi/30$ rad

(5) $2\pi/3$ rad

(6) $49\pi/15$ rad

(7) $107\pi/60$ rad

(8) $121\pi/60$ rad

(9) $17\pi/30$ rad

(10) $7\pi/60$ rad

(11) $\pi/5$ rad

(12) $7\pi/20$ rad

Instructions: Convert the given angle from radians to degrees. Check your answers in the back of the book.

(1) $\pi/60$ rad

(2) $17\pi/30$ rad

(3) $\pi/20$ rad

(4) $9\pi/2$ rad

(5) $\pi/2$ rad

(6) $43\pi/60$ rad

(7) $19\pi/60$ rad

(8) $4\pi/5$ rad

(9) $3\pi/10$ rad

(10) $79\pi/30$ rad

(11) $8\pi/15$ rad

(12) $11\pi/12$ rad

Instructions: Convert the given angle from radians to degrees. Check your answers in the back of the book.

(1) $11\pi/30$ rad

(2) $\pi/4$ rad

(3) $8\pi/15$ rad

(4) $29\pi/60$ rad

(5) $19\pi/60$ rad

(6) $11\pi/20$ rad

(7) $39\pi/20$ rad

(8) $2\pi/15$ rad

(9) $\pi/12$ rad

(10) $9\pi/10$ rad

(11) $9\pi/20$ rad

(12) $3\pi/10$ rad

Instructions: Convert the given angle from radians to degrees. Check your answers in the back of the book.

(1) $8\pi/5$ rad

(2) $13\pi/60$ rad

(3) $23\pi/5$ rad

(4) $31\pi/10$ rad

(5) $41\pi/20$ rad

(6) $14\pi/15$ rad

(7) $\pi/15$ rad

(8) $9\pi/20$ rad

(9) $47\pi/60$ rad

(10) $49\pi/60$ rad

(11) $29\pi/30$ rad

(12) $29\pi/60$ rad

Instructions: Convert the given angle from radians to degrees. Check your answers in the back of the book.

(1) $7\pi/20$ rad

(2) $119\pi/30$ rad

(3) $13\pi/60$ rad

(4) $13\pi/30$ rad

(5) $37\pi/60$ rad

(6) $5\pi/12$ rad

(7) $47\pi/60$ rad

(8) $41\pi/60$ rad

(9) $3\pi/20$ rad

(10) $13\pi/15$ rad

(11) $19\pi/30$ rad

(12) $13\pi/20$ rad

Instructions: Convert the given angle from radians to degrees. Check your answers in the back of the book.

(1) 227π/60 rad

(2) 9π/10 rad

(3) 47π/60 rad

(4) 3π/20 rad

(5) 11π/15 rad

(6) π/30 rad

(7) 113π/30 rad

(8) 13π/20 rad

(9) 2π/15 rad

(10) 13π/15 rad

(11) 71π/30 rad

(12) 7π/60 rad

Instructions: Convert the given angle from radians to degrees. Check your answers in the back of the book.

(1) $31\pi/20$ rad

(2) π rad

(3) $31\pi/60$ rad

(4) $13\pi/12$ rad

(5) 3π rad

(6) $5\pi/6$ rad

(7) $49\pi/60$ rad

(8) $79\pi/30$ rad

(9) $3\pi/10$ rad

(10) $37\pi/60$ rad

(11) $4\pi/5$ rad

(12) $29\pi/60$ rad

Instructions: Convert the given angle from radians to degrees. Check your answers in the back of the book.

(1) 49π/12 rad

(2) 19π/30 rad

(3) 3π/20 rad

(4) 8π/15 rad

(5) π/15 rad

(6) 8π/3 rad

(7) 4π/15 rad

(8) 11π/30 rad

(9) 2π/3 rad

(10) 3π/10 rad

(11) 3π/5 rad

(12) 62π/15 rad

Instructions: Convert the given angle from radians to degrees. Check your answers in the back of the book.

(1) $7\pi/12$ rad

(2) $21\pi/10$ rad

(3) $7\pi/6$ rad

(4) $7\pi/60$ rad

(5) $69\pi/20$ rad

(6) $217\pi/60$ rad

(7) $5\pi/6$ rad

(8) $227\pi/60$ rad

(9) $47\pi/12$ rad

(10) $\pi/6$ rad

(11) $7\pi/20$ rad

(12) $9\pi/20$ rad

Instructions: Convert the given angle from radians to degrees. Check your answers in the back of the book.

(1) $11\pi/15$ rad

(2) $11\pi/12$ rad

(3) $9\pi/10$ rad

(4) $89\pi/30$ rad

(5) $19\pi/60$ rad

(6) $\pi/3$ rad

(7) $41\pi/60$ rad

(8) $5\pi/2$ rad

(9) $8\pi/15$ rad

(10) $41\pi/30$ rad

(11) $47\pi/60$ rad

(12) $7\pi/15$ rad

Instructions: Convert the given angle from radians to degrees. Check your answers in the back of the book.

(1) 29π/6 rad

(2) 11π/12 rad

(3) 163π/60 rad

(4) 7π/60 rad

(5) 29π/10 rad

(6) 8π/15 rad

(7) π/3 rad

(8) 37π/60 rad

(9) 47π/60 rad

(10) 19π/60 rad

(11) 7π/10 rad

(12) 4π/15 rad

Chapter 3: Identifying Trig Functions in Right Triangles

The basic trig functions – sine (sin), cosine (cos), and tangent (tan) – are defined as the ratios of sides of right triangles. In a right triangle, the sine of an angle equals the ratio of the opposite side to the hypotenuse, the cosine of an angle equals the ratio of the adjacent side to the hypotenuse, and the tangent equals the ratio of the opposite side to the adjacent side.

A right triangle has one 90° angle. The opposite and adjacent sides make the 90° angle. The third side, which is the longest side, is called the hypotenuse. The angle in the argument of the trig function lies between the adjacent side and the hypotenuse. Study the triangle below. Also study the definitions of the sine, cosine, and tangent. Given different shapes and orientations of right triangles, it is crucial in trigonometry that you can correctly identify the sine, cosine, and tangent of a specified angle. This chapter provides extensive practice in order to help you develop this skill.

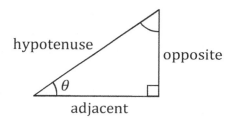

$$\sin \theta = \frac{\text{opposite}}{\text{hypotenuse}} \quad , \quad \cos \theta = \frac{\text{adjacent}}{\text{hypotenuse}} \quad , \quad \tan \theta = \frac{\text{opposite}}{\text{adjacent}}$$

It may be helpful to memorize soh-cah-toa, which is an abbreviation for sine-opposite-hypotenuse; cosine-adjacent-hypotenuse; tangent-opposite-adjacent. Students who can write down soh-cah-toa can often correctly identify the sine, cosine, and tangent of an angle when looking at a right triangle. Note that θ is the lowercase Greek letter theta, and is very commonly used as the variable for an unknown angle.

There are three additional trig functions – the cosecant (csc), secant (cos), and cotangent (cot) – which are reciprocals of the sine, cosine, and tangent functions:

$$\csc \theta = \frac{\text{hypotenuse}}{\text{opposite}} \quad , \quad \sec \theta = \frac{\text{hypotenuse}}{\text{adjacent}} \quad , \quad \cot \theta = \frac{\text{adjacent}}{\text{opposite}}$$

It is important to note is that the cosecant is the reciprocal of sine function, while the secant is the reciprocal of the cosine function; many students do not find this association to be intuitive.

In this chapter, each problem gives you a right triangle with two of the sides known and one unknown angle identified as θ. For each triangle, you need to determine two basic trig functions – i.e. sine, cosine, tangent, cosecant, secant, and cotangent. In order to do this, you will need to be able to determine which side is adjacent to θ, which side is opposite to θ, and which side is the hypotenuse. You will also need to apply the Pythagorean theorem in order to determine the unknown side.

According to the Pythagorean theorem, the opposite (o), adjacent (a), and hypotenuse (h) are related by the following equation:

$$o^2 + a^2 = h^2$$

Take some time to understand these concepts, and study the examples. Once you understand the following examples, you are ready to practice the technique yourself. You may need to refer to the examples frequently as you begin, but should try to solve the exercises all by yourself once you get the hang of it. Be sure to check the answers at the back of the book to ensure that you are solving the problems correctly.

Instructions: For each right triangle, express the two specified trig functions as reduced fractions.

Procedure: First, determine the unknown side using the Pythagorean theorem (as in the examples that follow). You need to be able to determine which side is opposite to θ, which side is adjacent to θ, and which side is the hypotenuse. Use the definitions of the basic trig functions (on the previous page) in order to express each trig function as a fraction.

See if each fraction is reducible. That is, if the numerator and denominator are both evenly divisible by an integer greater than 1, then the fraction is reducible. Reduce a fraction by dividing both the numerator and denominator by the greatest common factor (as in the examples that follow).

Factor any perfect squares out of the squareroot (as in one of the examples that follow). For example, in $\sqrt{12}$, we can factor 12 as 3 times 4. Since 4 is a perfect square $(2 \times 2 = 4)$, we can write $\sqrt{12} = \sqrt{3 \times 4} = \sqrt{3}\sqrt{4} = 2\sqrt{3}$. Also, rationalize the denominator (as in one of the examples that follow). For example, $\frac{1}{\sqrt{2}} = \frac{1}{\sqrt{2}}\frac{\sqrt{2}}{\sqrt{2}} = \frac{\sqrt{2}}{2}$ and $\frac{6}{\sqrt{3}} = \frac{6}{\sqrt{3}}\frac{\sqrt{3}}{\sqrt{3}} = \frac{6\sqrt{3}}{3} = 2\sqrt{3}$.

Example 1: $\sin\theta\,,\cos\theta\,=\,?$

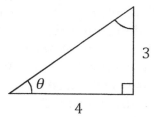

The adjacent to θ is 4, the opposite to θ is 3, and the hypotenuse can be found from the Pythagorean theorem: $h = \sqrt{a^2 + o^2} = \sqrt{4^2 + 3^2} = \sqrt{16+9} = \sqrt{25} = 5$.

$$\sin\theta = \frac{o}{h} = \frac{3}{5}\quad,\quad \cos\theta = \frac{a}{h} = \frac{4}{5}$$

Example 2: $\tan\theta\,,\cos\theta\,=\,?$

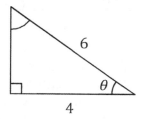

The adjacent to θ is 4, the hypotenuse is 6, and the opposite to θ can be found from the Pythagorean theorem: $o = \sqrt{h^2 - a^2} = \sqrt{6^2 - 4^2} = \sqrt{36-16} = \sqrt{20} = \sqrt{4}\sqrt{5} = 2\sqrt{5}$.

$$\tan\theta = \frac{o}{a} = \frac{2\sqrt{5}}{4} = \frac{\sqrt{5}}{2}\quad,\quad \cos\theta = \frac{a}{h} = \frac{4}{6} = \frac{2}{3}$$

Example 3: $\sin\theta\,,\sec\theta\,=\,?$

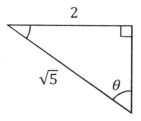

The opposite to θ is 2, the hypotenuse is $\sqrt{5}$, and the adjacent to θ can be found from the Pythagorean theorem: $a = \sqrt{h^2 - o^2} = \sqrt{\left(\sqrt{5}\right)^2 - 2^2} = \sqrt{5-4} = \sqrt{1} = 1$.

$$\sin\theta = \frac{o}{h} = \frac{2}{\sqrt{5}} = \frac{2}{\sqrt{5}}\frac{\sqrt{5}}{\sqrt{5}} = \frac{2\sqrt{5}}{5}\quad,\quad \sec\theta = \frac{h}{a} = \frac{\sqrt{5}}{1} = \sqrt{5}$$

Instructions: For each right triangle, express the two specified trig functions as reduced fractions. Check your answers in the back of the book.

Note: The triangles are <u>not drawn to scale</u>. The side that looks longer (except for the hypotenuse, which is always longest) may actually be shorter, and the angle that looks larger (except for 90°, which is clearly marked with a box) may actually be smaller.

(1) $\sin\theta$, $\cos\theta = ?$

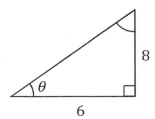

(2) $\sin\theta$, $\tan\theta = ?$

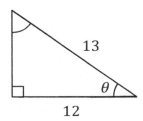

(3) $\cos\theta$, $\tan\theta = ?$

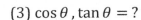
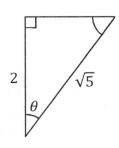

(4) $\cos\theta$, $\tan\theta = ?$

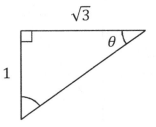

(5) $\sin\theta$, $\sec\theta = ?$

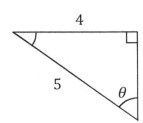

(6) $\csc\theta$, $\cos\theta = ?$

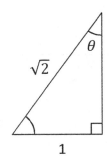

Instructions: For each right triangle, express the two specified trig functions as reduced fractions. Check your answers in the back of the book.

Note: The triangles are <u>not drawn to scale</u>. The side that looks longer (except for the hypotenuse, which is always longest) may actually be shorter, and the angle that looks larger (except for 90°, which is clearly marked with a box) may actually be smaller.

(1) $\cos\theta$, $\tan\theta = ?$

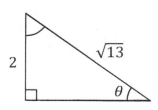

(2) $\sin\theta$, $\cos\theta = ?$

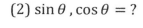
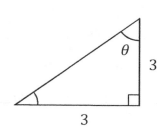

(3) $\sin\theta$, $\cot\theta = ?$

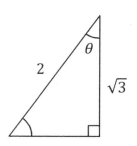

(4) $\cos\theta$, $\cot\theta = ?$

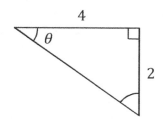

(5) $\csc\theta$, $\sec\theta = ?$

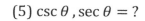
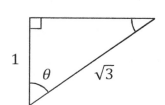

(6) $\csc\theta$, $\tan\theta = ?$

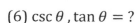
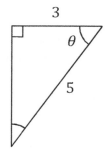

Instructions: For each right triangle, express the two specified trig functions as reduced fractions. Check your answers in the back of the book.

Note: The triangles are <u>not drawn to scale</u>. The side that looks longer (except for the hypotenuse, which is always longest) may actually be shorter, and the angle that looks larger (except for 90°, which is clearly marked with a box) may actually be smaller.

(1) $\sin\theta, \cos\theta = ?$

(2) $\sin\theta, \tan\theta = ?$

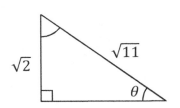

(3) $\cos\theta, \tan\theta = ?$

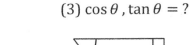

(4) $\cos\theta, \tan\theta = ?$

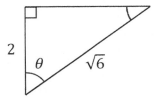

(5) $\sin\theta, \sec\theta = ?$

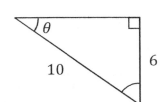

(6) $\csc\theta, \cos\theta = ?$

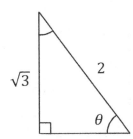

Instructions: For each right triangle, express the two specified trig functions as reduced fractions. Check your answers in the back of the book.

Note: The triangles are <u>not drawn to scale</u>. The side that looks longer (except for the hypotenuse, which is always longest) may actually be shorter, and the angle that looks larger (except for 90°, which is clearly marked with a box) may actually be smaller.

(1) $\sin\theta$, $\cot\theta = ?$

(2) $\cos\theta$, $\tan\theta = ?$

(3) $\sin\theta$, $\sec\theta = ?$

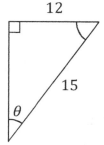

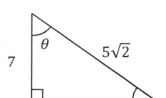

(4) $\cos\theta$, $\tan\theta = ?$

(5) $\sin\theta$, $\cos\theta = ?$

(6) $\csc\theta$, $\cot\theta = ?$

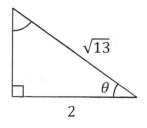

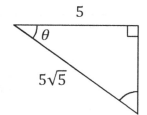

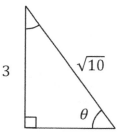

Instructions: For each right triangle, express the two specified trig functions as reduced fractions. Check your answers in the back of the book.

Note: The triangles are <u>not drawn to scale</u>. The side that looks longer (except for the hypotenuse, which is always longest) may actually be shorter, and the angle that looks larger (except for 90°, which is clearly marked with a box) may actually be smaller.

(1) $\sin\theta$, $\cos\theta$ = ?

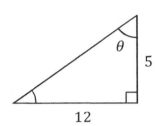

(2) $\sec\theta$, $\tan\theta$ = ?

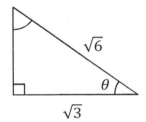

(3) $\csc\theta$, $\tan\theta$ = ?

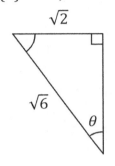

(4) $\sin\theta$, $\tan\theta$ = ?

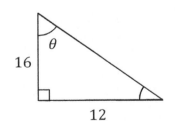

(5) $\sin\theta$, $\sec\theta$ = ?

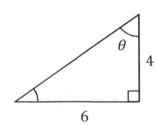

(6) $\cos\theta$, $\cot\theta$ = ?

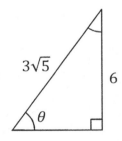

Instructions: For each right triangle, express the two specified trig functions as reduced fractions. Check your answers in the back of the book.

Note: The triangles are <u>not drawn to scale</u>. The side that looks longer (except for the hypotenuse, which is always longest) may actually be shorter, and the angle that looks larger (except for 90°, which is clearly marked with a box) may actually be smaller.

(1) $\sec\theta$, $\csc\theta = ?$

(2) $\sec\theta$, $\cot\theta = ?$

(3) $\csc\theta$, $\cot\theta = ?$

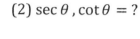

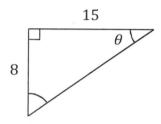

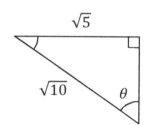

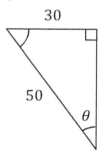

(4) $\sin\theta$, $\cot\theta = ?$

(5) $\cos\theta$, $\csc\theta = ?$

(6) $\csc\theta$, $\tan\theta = ?$

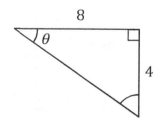

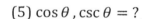

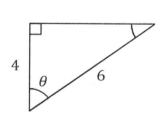

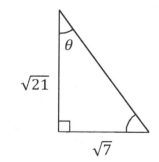

Instructions: For each right triangle, express the two specified trig functions as reduced fractions. Check your answers in the back of the book.

Note: The triangles are <u>not drawn to scale</u>. The side that looks longer (except for the hypotenuse, which is always longest) may actually be shorter, and the angle that looks larger (except for 90°, which is clearly marked with a box) may actually be smaller.

(1) $\sin\theta$, $\tan\theta = ?$

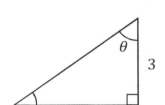

(2) $\sin\theta$, $\cos\theta = ?$

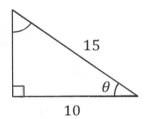

(3) $\sec\theta$, $\tan\theta = ?$

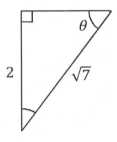

(4) $\cos\theta$, $\cot\theta = ?$

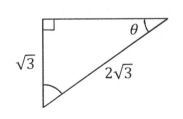

(5) $\sin\theta$, $\sec\theta = ?$

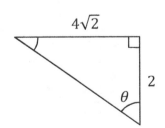

(6) $\csc\theta$, $\cos\theta = ?$

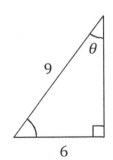

Instructions: For each right triangle, express the two specified trig functions as reduced fractions. Check your answers in the back of the book.

Note: The triangles are <u>not drawn to scale</u>. The side that looks longer (except for the hypotenuse, which is always longest) may actually be shorter, and the angle that looks larger (except for 90°, which is clearly marked with a box) may actually be smaller.

(1) $\sin\theta$, $\tan\theta = ?$

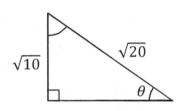

(2) $\sin\theta$, $\tan\theta = ?$

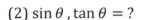
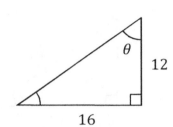

(3) $\cos\theta$, $\cot\theta = ?$

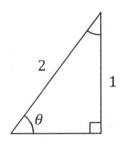

(4) $\sec\theta$, $\tan\theta = ?$

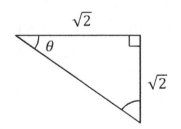

(5) $\csc\theta$, $\sec\theta = ?$

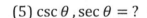
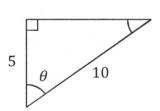

(6) $\csc\theta$, $\cot\theta = ?$

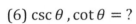
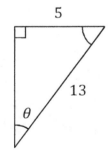

Instructions: For each right triangle, express the two specified trig functions as reduced fractions. Check your answers in the back of the book.

Note: The triangles are <u>not drawn to scale</u>. The side that looks longer (except for the hypotenuse, which is always longest) may actually be shorter, and the angle that looks larger (except for 90°, which is clearly marked with a box) may actually be smaller.

(1) $\csc\theta$, $\cos\theta = ?$

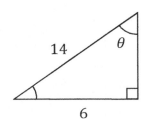

(2) $\sin\theta$, $\sec\theta = ?$

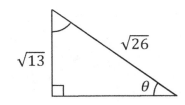

(3) $\cos\theta$, $\tan\theta = ?$

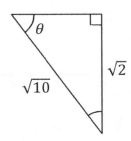

(4) $\cos\theta$, $\tan\theta = ?$

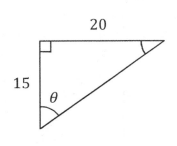

(5) $\sin\theta$, $\sec\theta = ?$

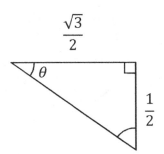

(6) $\sin\theta$, $\cos\theta = ?$

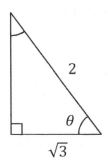

Instructions: For each right triangle, express the two specified trig functions as reduced fractions. Check your answers in the back of the book.

Note: The triangles are <u>not drawn to scale</u>. The side that looks longer (except for the hypotenuse, which is always longest) may actually be shorter, and the angle that looks larger (except for 90°, which is clearly marked with a box) may actually be smaller.

(1) $\cos\theta$, $\cot\theta = ?$

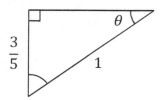

(2) $\sin\theta$, $\tan\theta = ?$

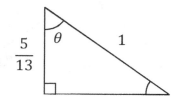

(3) $\sin\theta$, $\sec\theta = ?$

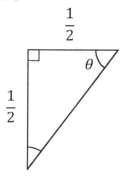

(4) $\cos\theta$, $\cot\theta = ?$

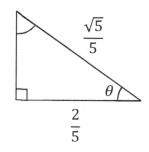

(5) $\sin\theta$, $\cos\theta = ?$

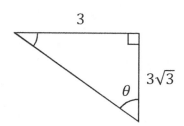

(6) $\sec\theta$, $\cot\theta = ?$

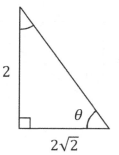

Instructions: For each right triangle, express the two specified trig functions as reduced fractions. Check your answers in the back of the book.

Note: The triangles are <u>not drawn to scale</u>. The side that looks longer (except for the hypotenuse, which is always longest) may actually be shorter, and the angle that looks larger (except for 90°, which is clearly marked with a box) may actually be smaller.

(1) $\sin\theta$, $\cos\theta = ?$

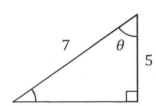

(2) $\sec\theta$, $\tan\theta = ?$

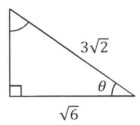

(3) $\csc\theta$, $\tan\theta = ?$

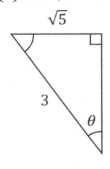

(4) $\sin\theta$, $\tan\theta = ?$

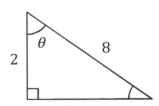

(5) $\sin\theta$, $\sec\theta = ?$

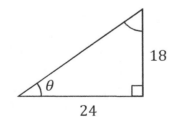

(6) $\cos\theta$, $\cot\theta = ?$

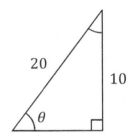

Instructions: For each right triangle, express the two specified trig functions as reduced fractions. Check your answers in the back of the book.

Note: The triangles are <u>not drawn to scale</u>. The side that looks longer (except for the hypotenuse, which is always longest) may actually be shorter, and the angle that looks larger (except for 90°, which is clearly marked with a box) may actually be smaller.

(1) $\sec\theta$, $\csc\theta$ = ?

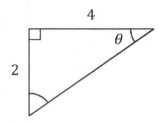

(2) $\sec\theta$, $\cot\theta$ = ?

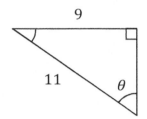

(3) $\csc\theta$, $\cot\theta$ = ?

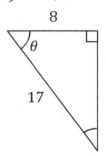

(4) $\sin\theta$, $\cot\theta$ = ?

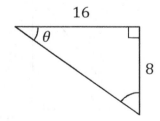

(5) $\cos\theta$, $\csc\theta$ = ?

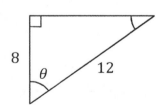

(6) $\sin\theta$, $\cos\theta$ = ?

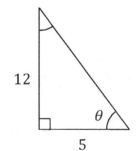

Chapter 4: Special Right Triangles

Two special right triangles that are very common are the 30°-60°-90° and 45°-45°-90° triangles. These right triangles are encountered in math and science courses – and standardized math exams – so frequently that it is advantageous to be able to recognize these triangles and to be able to fluently solve for any unknown angle or side in these triangles.

The 30°-60°-90° triangle is one-half of an equilateral triangle. Therefore, the side opposite to 30° is half as long as the hypotenuse. If the side opposite to 30° is 1, the hypotenuse is 2. The remaining side, which is adjacent to 30° (but opposite to 60°), can be found from the Pythagorean theorem: $a = \sqrt{h^2 - o^2} = \sqrt{2^2 - 1^2} = \sqrt{4 - 1} = \sqrt{3}$. In general, we can say that the ratio of the sides of a 30°-60°-90° triangle is $1 : \sqrt{3} : 2$, where 1 is opposite to 30°, $\sqrt{3}$ is opposite to 60°, and 2 is the hypotenuse. This is worth memorizing.

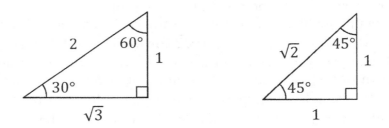

The 45°-45°-90° has two equal sides, opposite to the two equal angles. If these sides are each 1, the hypotenuse is $h = \sqrt{a^2 + o^2} = \sqrt{1^2 + 1^2} = \sqrt{1 + 1} = \sqrt{2}$, according to the Pythagorean theorem. In general, we can say that the ratio of the sides of a 45°-45°-90° triangle is $1 : 1 : \sqrt{2}$, where each 1 is opposite to a 45° angle and $\sqrt{2}$ is the hypotenuse. This is also worth memorizing.

A triangle has the same angles if you multiply all three sides by the same factor. For example, a triangle with sides $2 : 2\sqrt{3} : 4$ is a 30°-60°-90° triangle, and a triangle with sides $3 : 3 : 3\sqrt{2}$ is a 45°-45°-90° triangle.

If you know one side and one angle or if you know two sides of a 30°-60°-90° or 45°-45°-90° triangle, you can solve for the remaining sides by knowing the ratios $1 : \sqrt{3} : 2$ and $1 : 1 : \sqrt{2}$, respectively; and you can also determine the unknown angle(s) by remembering which side is the hypotenuse and which angle is opposite to each side. This technique is illustrated in the following examples.

In a 30°-60°-90° triangle, the shortest side is always opposite to 30° and the longest side is always the hypotenuse. Also, the side opposite to 60° is always $\sqrt{3}$ times longer than the side opposite to 30°, and the hypotenuse is always 2 times longer than the side opposite to 30°.

In a 45°-45°-90° triangle, the longest side is always the hypotenuse and the other two sides are equal. Also, the hypotenuse is always $\sqrt{2}$ times longer than the other sides.

Take some time to understand these concepts, and study the examples. Once you understand the following examples, you are ready to practice the technique yourself. You may need to refer to the examples frequently as you begin, but should try to solve the exercises all by yourself once you get the hang of it. Be sure to check the answers at the back of the book to ensure that you are solving the problems correctly.

Instructions: For each right triangle, determine the indicated unknown distances and/or angles. An unknown angle will be called θ, while an unknown distance will be called x or y.

Procedure: First determine whether the triangle is a 30°-60°-90° or 45°-45°-90° triangle. If two sides are equal, it's a 45°-45°-90° triangle.

If either angle is given in the problem (in addition to the right angle), you should easily determine the missing angle. If neither angle is given, but two of the sides are equal, the two angles are 45°. If neither angle is given, and none of the sides are equal, the angle opposite to the shortest side is 30° and the other angle is 60°.

Determine the unknown side(s) using the ratio of the triangle. The ratio of the sides is $1:\sqrt{3}:2$ for a 30°-60°-90° triangle and $1:1:\sqrt{2}$ for a 45°-45°-90° triangle. Multiply the known side(s) by the appropriate factor – using the known ratio of the sides – in order to solve for the unknown side(s). This technique is illustrated in the following examples.

In a 45°-45°-90° triangle, the hypotenuse is $\sqrt{2}$ times longer than the other sides. In a 30°-60°-90° triangle, the side opposite to 60° is $\sqrt{3}$ times longer than the side opposite to 30°, and the hypotenuse is 2 times longer than the side opposite to 30°.

Reduce any fractions by dividing both the numerator and denominator by the greatest common factor. Factor any perfect squares out of the squareroot. For example, in $\sqrt{12}$, we can factor 12 as 3 times 4. Since 4 is a perfect square ($2 \times 2 = 4$), we can write $\sqrt{12} = \sqrt{3 \times 4} = \sqrt{3}\sqrt{4} = 2\sqrt{3}$. Also, rationalize the denominator. For example, $\frac{1}{\sqrt{2}} = \frac{1}{\sqrt{2}}\frac{\sqrt{2}}{\sqrt{2}} = \frac{\sqrt{2}}{2}$ and $\frac{6}{\sqrt{3}} = \frac{6}{\sqrt{3}}\frac{\sqrt{3}}{\sqrt{3}} = \frac{6\sqrt{3}}{3} = 2\sqrt{3}$.

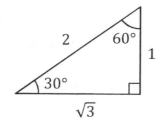

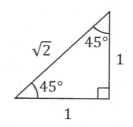

Example 1: $x, y = ?$

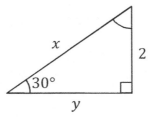

This is a 30°-60°-90° triangle. The variable x is on the hypotenuse, and is 2 times longer than the side opposite of 30°: $x = 2 \times 2 = 4$. The variable y is opposite of 60° (the angle that was not labeled), and is $\sqrt{3}$ times longer than the side opposite of 30°: $y = \sqrt{3} \times 2 = 2\sqrt{3}$.

Example 2: $\theta, x = ?$

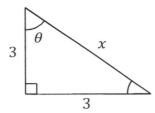

This is a 45°-45°-90° triangle because two sides are equal. Therefore, $\theta = 45°$. The variable x is on the hypotenuse, and is $\sqrt{2}$ times longer than the other sides: $x = \sqrt{2} \times 3 = 3\sqrt{2}$.

Example 3: $x, y = ?$

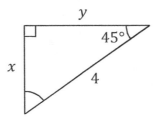

This is a 45°-45°-90° triangle. The hypotenuse is 4. The two sides, x and y, are $\sqrt{2}$ times shorter than the hypotenuse:

$$x = \frac{4}{\sqrt{2}} = \frac{4}{\sqrt{2}}\frac{\sqrt{2}}{\sqrt{2}} = \frac{4\sqrt{2}}{2} = 2\sqrt{2}$$
$$y = 2\sqrt{2}$$

Example 4: $x, y = ?$

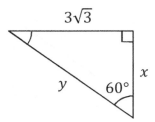

This is a 30°-60°-90° triangle. The side opposite of 60° is $3\sqrt{3}$. The variable x is opposite of 30° (the angle that was not labeled), and is $\sqrt{3}$ times shorter than the side opposite to 60°: $x = \frac{3\sqrt{3}}{\sqrt{3}} = 3$. The variable y is on the hypotenuse, and is 2 times longer than the side opposite of 30°: $y = 2x = 2 \times 3 = 6$.

Example 5: $\theta, x = ?$

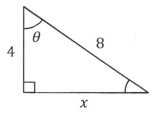

This is a 30°-60°-90° triangle because the hypotenuse is twice as long as one of the sides. Since 4 is one-half of the hypotenuse, the angle that was not labeled is 30°, so the labeled angle is $\theta = 60°$. The variable x is opposite of 60°, and is $\sqrt{3}$ times longer than the side opposite of 30°: $x = \sqrt{3} \times 4 = 4\sqrt{3}$.

Example 6: $x, y = ?$

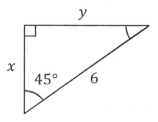

This is a 45°-45°-90° triangle. The hypotenuse is 6. The two sides, x and y, are $\sqrt{2}$ times shorter than the hypotenuse:

$$x = \frac{6}{\sqrt{2}} = \frac{6}{\sqrt{2}}\frac{\sqrt{2}}{\sqrt{2}} = \frac{6\sqrt{2}}{2} = 3\sqrt{2}$$
$$y = 3\sqrt{2}$$

Instructions: For each right triangle, determine the unknown distance(s) and angle(s). Check your answers in the back of the book.

Note: The triangles are <u>not drawn to scale</u>. The side that looks longer (except for the hypotenuse, which is always longest) may actually be shorter, and the angle that looks larger (except for 90°, which is clearly marked with a box) may actually be smaller.

(1) $x, y = ?$

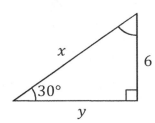

(2) $\theta, x = ?$

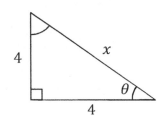

(3) $x, y = ?$

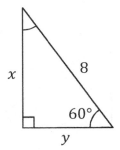

(4) $x, y = ?$

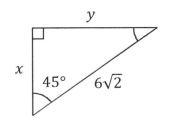

(5) $\theta, x = ?$

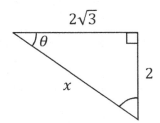

(6) $x, y = ?$

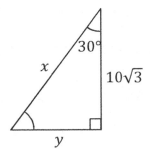

Instructions: For each right triangle, determine the unknown distance(s) and angle(s). Check your answers in the back of the book.

Note: The triangles are not drawn to scale. The side that looks longer (except for the hypotenuse, which is always longest) may actually be shorter, and the angle that looks larger (except for 90°, which is clearly marked with a box) may actually be smaller.

(1) $\theta, x = ?$

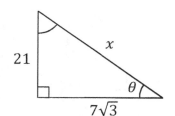

(2) $x, y = ?$

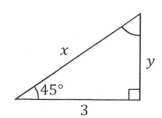

(3) $x, y = ?$

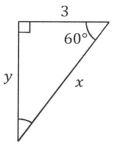

(4) $\theta, x = ?$

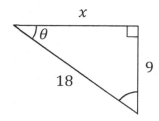

(5) $\theta, x = ?$

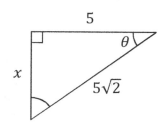

(6) $x, y = ?$

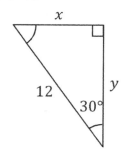

Instructions: For each right triangle, determine the unknown distance(s) and angle(s). Check your answers in the back of the book.

Note: The triangles are <u>not drawn to scale</u>. The side that looks longer (except for the hypotenuse, which is always longest) may actually be shorter, and the angle that looks larger (except for 90°, which is clearly marked with a box) may actually be smaller.

(1) $x, y = ?$

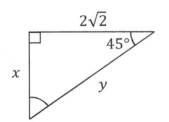

(2) $x, y = ?$

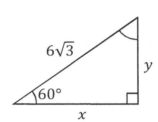

(3) $x, y = ?$

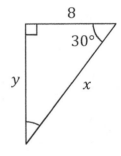

(4) $x, y = ?$

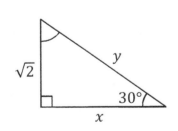

(5) $\theta, x = ?$

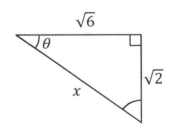

(6) $x, y = ?$

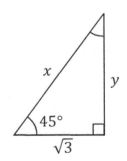

Instructions: For each right triangle, determine the unknown distance(s) and angle(s). Check your answers in the back of the book.

Note: The triangles are <u>not drawn to scale</u>. The side that looks longer (except for the hypotenuse, which is always longest) may actually be shorter, and the angle that looks larger (except for 90°, which is clearly marked with a box) may actually be smaller.

(1) $x, y = ?$

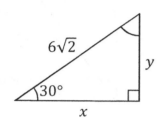

(2) $x, y = ?$

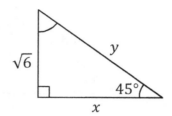

(3) $\theta, x = ?$

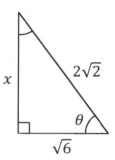

(4) $x, y = ?$

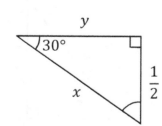

(5) $x, y = ?$

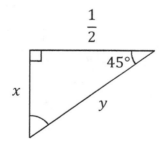

(6) $x, y = ?$

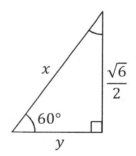

Instructions: For each right triangle, determine the unknown distance(s) and angle(s). Check your answers in the back of the book.

Note: The triangles are <u>not drawn to scale</u>. The side that looks longer (except for the hypotenuse, which is always longest) may actually be shorter, and the angle that looks larger (except for 90°, which is clearly marked with a box) may actually be smaller.

(1) $\theta, x = ?$

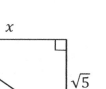

(2) $\theta, x = ?$

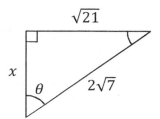

(3) $x, y = ?$

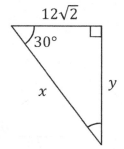

(4) $x, y = ?$

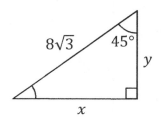

(5) $\theta, x = ?$

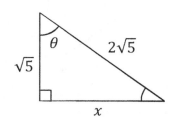

(6) $\theta, x = ?$

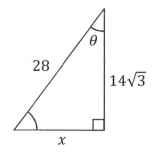

Instructions: For each right triangle, determine the unknown distance(s) and angle(s). Check your answers in the back of the book.

Note: The triangles are <u>not drawn to scale</u>. The side that looks longer (except for the hypotenuse, which is always longest) may actually be shorter, and the angle that looks larger (except for 90°, which is clearly marked with a box) may actually be smaller.

(1) $\theta, x = ?$

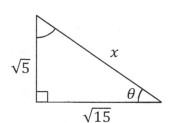

(2) $\theta, x = ?$

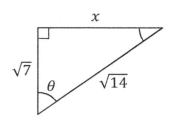

(3) $x, y = ?$

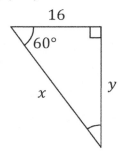

(4) $\theta, x = ?$

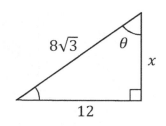

(5) $\theta, x = ?$

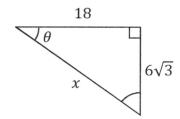

(6) $x, y = ?$

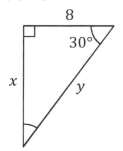

Instructions: For each right triangle, determine the unknown distance(s) and angle(s). Check your answers in the back of the book.

Note: The triangles are not drawn to scale. The side that looks longer (except for the hypotenuse, which is always longest) may actually be shorter, and the angle that looks larger (except for 90°, which is clearly marked with a box) may actually be smaller.

(1) $x, y = ?$

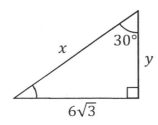

(2) $\theta, x = ?$

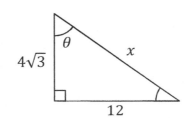

(3) $x, y = ?$

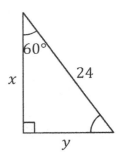

(4) $x, y = ?$

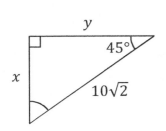

(5) $\theta, x = ?$

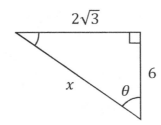

(6) $\theta, x = ?$

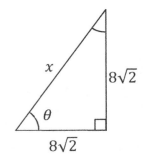

Instructions: For each right triangle, determine the unknown distance(s) and angle(s). Check your answers in the back of the book.

Note: The triangles are not drawn to scale. The side that looks longer (except for the hypotenuse, which is always longest) may actually be shorter, and the angle that looks larger (except for 90°, which is clearly marked with a box) may actually be smaller.

(1) $\theta, x = ?$

(2) $x, y = ?$

(3) $x, y = ?$

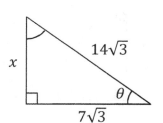

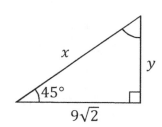

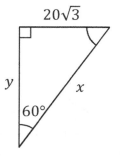

(4) $\theta, x = ?$

(5) $\theta, x = ?$

(6) $x, y = ?$

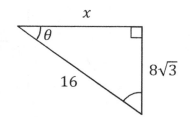

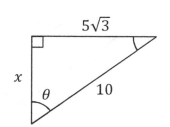

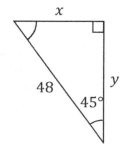

Instructions: For each right triangle, determine the unknown distance(s) and angle(s). Check your answers in the back of the book.

Note: The triangles are <u>not drawn to scale</u>. The side that looks longer (except for the hypotenuse, which is always longest) may actually be shorter, and the angle that looks larger (except for 90°, which is clearly marked with a box) may actually be smaller.

(1) $x, y = ?$

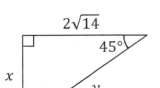

(2) $x, y = ?$

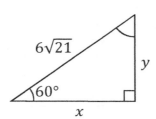

(3) $x, y = ?$

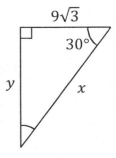

(4) $x, y = ?$

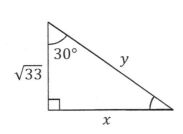

(5) $\theta, x = ?$

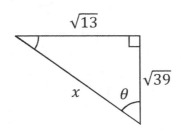

(6) $x, y = ?$

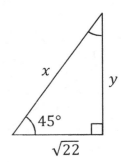

Instructions: For each right triangle, determine the unknown distance(s) and angle(s). Check your answers in the back of the book.

Note: The triangles are <u>not drawn to scale</u>. The side that looks longer (except for the hypotenuse, which is always longest) may actually be shorter, and the angle that looks larger (except for 90°, which is clearly marked with a box) may actually be smaller.

(1) $x, y = ?$

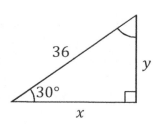

(2) $x, y = ?$

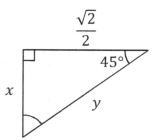

(3) $\theta, x = ?$

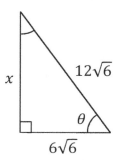

(4) $x, y = ?$

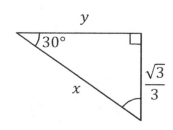

(5) $x, y = ?$

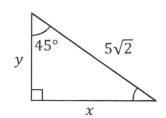

(6) $x, y = ?$

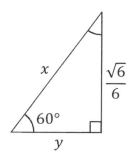

Instructions: For each right triangle, determine the unknown distance(s) and angle(s). Check your answers in the back of the book.

Note: The triangles are <u>not drawn to scale</u>. The side that looks longer (except for the hypotenuse, which is always longest) may actually be shorter, and the angle that looks larger (except for 90°, which is clearly marked with a box) may actually be smaller.

(1) $\theta, x = ?$

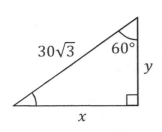

(2) $\theta, x = ?$

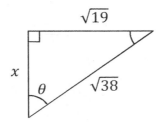

(3) $x, y = ?$

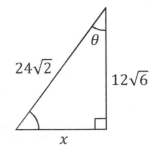

(4) $x, y = ?$

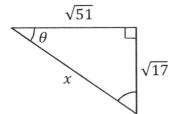

(5) $\theta, x = ?$

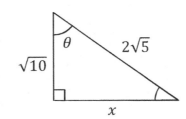

(6) $\theta, x = ?$

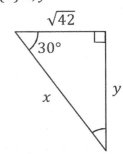

Instructions: For each right triangle, determine the unknown distance(s) and angle(s). Check your answers in the back of the book.

Note: The triangles are <u>not drawn to scale</u>. The side that looks longer (except for the hypotenuse, which is always longest) may actually be shorter, and the angle that looks larger (except for 90°, which is clearly marked with a box) may actually be smaller.

(1) $\theta, x = ?$

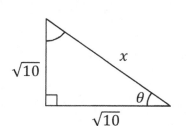

(2) $\theta, x = ?$

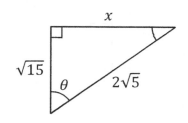

(3) $x, y = ?$

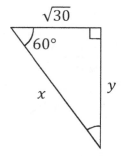

(4) $\theta, x = ?$

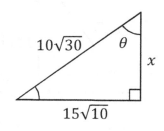

(5) $\theta, x = ?$

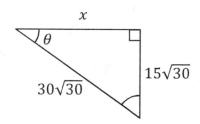

(6) $x, y = ?$

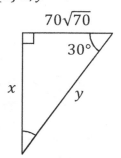

Chapter 5: Memorize Basic Trig Functions in Quadrant I

It is very useful to know the sine, cosine, and tangent of 0°, 30°, 45°, 60°, and 90°. These angles are used frequently in many math, physics, and standardized test problems since they don't require use of a calculator. Many trigonometry and physics teachers require their students to memorize this. The exercises of this chapter provide ample practice to develop fluency with the basic trig functions evaluated at these angles. Before we begin with the memorization exercises, we will derive each trig function's value at each of these angles.

In the 45°-45°-90° triangle illustrated below, the opposite and adjacent to either 45° angle are both equal to 1, while the hypotenuse is $\sqrt{2}$. Therefore, the sine, cosine, and tangent of 45° are:

$$\sin 45° = \frac{\text{opposite}}{\text{hypotenuse}} = \frac{1}{\sqrt{2}} = \frac{1}{\sqrt{2}}\frac{\sqrt{2}}{\sqrt{2}} = \frac{\sqrt{2}}{2}$$

$$\cos 45° = \frac{\text{adjacent}}{\text{hypotenuse}} = \frac{1}{\sqrt{2}} = \frac{1}{\sqrt{2}}\frac{\sqrt{2}}{\sqrt{2}} = \frac{\sqrt{2}}{2}$$

$$\tan 45° = \frac{\text{opposite}}{\text{adjacent}} = \frac{1}{1} = 1$$

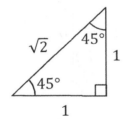
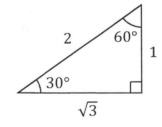

In the 30°-60°-90° triangle illustrated above, the side opposite to 30° is 1, the side adjacent to 30° is $\sqrt{3}$, and the hypotenuse is 2. Therefore, the sine, cosine, and tangent of 30° are:

$$\sin 30° = \frac{\text{opposite}}{\text{hypotenuse}} = \frac{1}{2}$$

$$\cos 30° = \frac{\text{adjacent}}{\text{hypotenuse}} = \frac{\sqrt{3}}{2}$$

$$\tan 30° = \frac{\text{opposite}}{\text{adjacent}} = \frac{1}{\sqrt{3}} = \frac{1}{\sqrt{3}}\frac{\sqrt{3}}{\sqrt{3}} = \frac{\sqrt{3}}{3}$$

Looking at the 60° angle, the side opposite to 60° is $\sqrt{3}$, the side adjacent to 60° is 1, and the hypotenuse is 2. Therefore, the sine, cosine, and tangent of 60° are:

$$\sin 60° = \frac{\text{opposite}}{\text{hypotenuse}} = \frac{\sqrt{3}}{2}$$

$$\cos 60° = \frac{\text{adjacent}}{\text{hypotenuse}} = \frac{1}{2}$$

$$\tan 60° = \frac{\text{opposite}}{\text{adjacent}} = \frac{\sqrt{3}}{1} = \sqrt{3}$$

As θ becomes smaller, the opposite side shrinks toward zero and the adjacent side grows longer, as illustrated below. In the limit that θ approaches zero, the opposite side becomes zero and the adjacent side becomes equal to the hypotenuse:

$$\sin 0° = \frac{\text{opposite}}{\text{hypotenuse}} = 0$$

$$\cos 0° = \frac{\text{adjacent}}{\text{hypotenuse}} = 1$$

$$\tan 0° = \frac{\text{opposite}}{\text{adjacent}} = 0$$

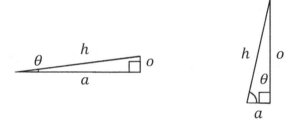

As θ becomes larger, the adjacent side shrinks toward zero and the opposite side grows longer, as illustrated above. In the limit that θ approaches 90°, the adjacent side becomes zero and the opposite side becomes equal to the hypotenuse:

$$\sin 90° = \frac{\text{opposite}}{\text{hypotenuse}} = 1$$

$$\cos 90° = \frac{\text{adjacent}}{\text{hypotenuse}} = 0$$

$$\tan 90° = \frac{\text{opposite}}{\text{adjacent}} = \text{undefined}$$

The tangent of 90° is undefined because of the division by zero (the opposite side is finite, while the adjacent side is zero).

All of the values of the trig functions that we just derived are tabulated below.

θ	0°	30°	45°	60°	90°
$\sin\theta$	0	$\dfrac{1}{2}$	$\dfrac{\sqrt{2}}{2}$	$\dfrac{\sqrt{3}}{2}$	1
$\cos\theta$	1	$\dfrac{\sqrt{3}}{2}$	$\dfrac{\sqrt{2}}{2}$	$\dfrac{1}{2}$	0
$\tan\theta$	0	$\dfrac{\sqrt{3}}{3}$	1	$\sqrt{3}$	undefined

Instead of also memorizing the values of the secant, cosecant, and cotangent functions at these angles, it is simpler just to memorize the following reciprocal relationships:

$$\csc\theta = \frac{1}{\sin\theta} \quad , \quad \sec\theta = \frac{1}{\cos\theta} \quad , \quad \cot\theta = \frac{1}{\tan\theta}$$

For example, if you know that the sine of 30° is 1/2 and that the cosecant function is the reciprocal of the sine function, then you can easily determine that $\csc 30° = 2$.

Memorization technique: Here is a simple method for memorizing the above table:

(1) First, write down the integers 0 thru 4 in order.

0	1	2	3	4

(2) Next, squareroot each number.

0	1	$\sqrt{2}$	$\sqrt{3}$	2

(3) Now divide each number by 2. These are the sines of 0°, 30°, 45°, 60°, and 90°.

0	1/2	$\sqrt{2}/2$	$\sqrt{3}/2$	1

(4) To get the cosines of the same angles, write the same numbers in reverse.

1	$\sqrt{3}/2$	$\sqrt{2}/2$	1/2	0

(5) To get the tangents of the same angles, divide the sine by the cosine.

0	$\sqrt{3}/3$	1	$\sqrt{3}$	undefined

Take some time to try to memorize the sine and cosine of 0°, 30°, 45°, 60°, and 90°. It may help to make flashcards with the problem on one side and the corresponding answer on the other side. You also need to memorize that the tangent equals the sine divided by the cosine, and how to obtain the cosecant, secant, and cotangent from the sine, cosine, and tangent.

$$\tan \theta = \frac{\sin \theta}{\cos \theta} \quad , \quad \csc \theta = \frac{1}{\sin \theta} \quad , \quad \sec \theta = \frac{1}{\cos \theta} \quad , \quad \cot \theta = \frac{1}{\tan \theta}$$

Instructions: Determine the answer to each problem. You should know the answers to the sines and cosines after memorizing them. You can determine the answer to a tangent by dividing sine by cosine. The secant, cosecant, and cotangent can be found by finding the reciprocal of the cosine, sine, and tangent, respectively. Some of the angles are given in degrees, while some angles are given in radians. Degrees will be indicated with a degree symbol (°), whereas angles in radians will have a π, but the radians will be implied.

Procedure: First, memorize the sine and cosine of 0°, 30°, 45°, 60°, and 90° (or memorize how to make the table described at the bottom of the previous page). If the problem is a sine or cosine, you will just have an answer, but will not have any work to show (unless you need to first convert from radians to degrees). If the problem is a tangent, divide the sine of the angle by the cosine of the angle. If the problem is a secant, cosecant, or cotangent, take a reciprocal of the cosine, sine, or tangent, respectively. If the angle is given in radians instead of degrees, you can convert the angle to degrees using the method of Chapter 2.

Rationalize the denominator. For example, $\frac{1}{\sqrt{2}} = \frac{1}{\sqrt{2}}\frac{\sqrt{2}}{\sqrt{2}} = \frac{\sqrt{2}}{2}$ and $\frac{1}{\sqrt{3}} = \frac{1}{\sqrt{3}}\frac{\sqrt{3}}{\sqrt{3}} = \frac{\sqrt{3}}{3}$.

Example 1: $\sin 60°$. According to the table, the answer is $\sqrt{3}/2$.

Example 2: $\cos(\pi/4)$. First convert the angle from radians to degrees:
$$\frac{\pi}{4} \text{ rad} \times \frac{180°}{\pi \text{ rad}} = \frac{180°}{4} = 45°$$
According to the table, the cosine of 45° is $\sqrt{2}/2$.

Example 3: $\tan 30°$. The formula states to divide $\sin 30°$ by $\cos 30°$:
$$\tan 30° = \frac{\sin 30°}{\cos 30°} = \frac{1/2}{\sqrt{3}/2} = \frac{1}{\sqrt{3}} = \frac{1}{\sqrt{3}}\frac{\sqrt{3}}{\sqrt{3}} = \frac{\sqrt{3}}{3}$$
To divide two fractions, multiply the first fraction by the reciprocal of the second fraction.

Example 3: $\sec 90°$. The formula states to divide 1 by $\cos 90°$.
$$\sec 90° = \frac{1}{\cos 90°} = \frac{1}{0} = \text{undefined}$$

Instructions: Determine the answer to each problem. Check your answers in the back of the book.

(1) sec 30°

(2) sin 30°

(3) cos(π/4)

(4) csc 0°

(5) sin(π/6)

(6) tan(π/3)

(7) tan 0°

(8) sin 30°

(9) sin 0°

(10) sec 45°

(11) cot 60°

(12) sec 90°

(13) tan 90°

(14) sec 30°

(15) csc 30°

(16) sin(π/3)

(17) sec(π/4)

(18) sin 30°

(19) cos 45°

(20) sec 0°

(21) csc 90°

(22) sec 60°

(23) cos 90°

(24) cos 0°

Instructions: Determine the answer to each problem. Check your answers in the back of the book.

(1) csc 45°

(2) sec(π/3)

(3) csc(π/6)

(4) sin(π/2)

(5) csc(π/4)

(6) csc 45°

(7) tan 90°

(8) sec 45°

(9) sin(π/6)

(10) csc 0°

(11) cos(π/2)

(12) tan 90°

(13) cot 45°

(14) sec 0°

(15) cos 0°

(16) cos 60°

(17) tan 90°

(18) cos(π/3)

(19) tan 0°

(20) cot 0°

(21) sin 30°

(22) sin(π/6)

(23) cos 45°

(24) cos 90°

Instructions: Determine the answer to each problem. Check your answers in the back of the book.

(1) $\cos 0°$

(2) $\tan(\pi/2)$

(3) $\cot 45°$

(4) $\sec 0°$

(5) $\sec 30°$

(6) $\cot 0°$

(7) $\csc(\pi/4)$

(8) $\cot 90°$

(9) $\sec 90°$

(10) $\sin(\pi/3)$

(11) $\csc 30°$

(12) $\sin 60°$

(13) $\cos(\pi/3)$

(14) $\cot 0°$

(15) $\cot 30°$

(16) $\cot 60°$

(17) $\csc(\pi/3)$

(18) $\sec 60°$

(19) $\sin(\pi/2)$

(20) $\cot 0°$

(21) $\sec 0°$

(22) $\sec(\pi/6)$

(23) $\cos(\pi/2)$

(24) $\sec 30°$

Instructions: Determine the answer to each problem. Check your answers in the back of the book.

(1) $\sin 30°$

(2) $\sec 0°$

(3) $\tan(\pi/6)$

(4) $\cos 60°$

(5) $\csc 90°$

(6) $\cot(\pi/2)$

(7) $\tan 45°$

(8) $\sec 30°$

(9) $\sec(\pi/3)$

(10) $\csc 30°$

(11) $\cot 45°$

(12) $\cos(\pi/4)$

(13) $\sin 0°$

(14) $\tan(\pi/2)$

(15) $\csc(\pi/4)$

(16) $\cot 90°$

(17) $\tan 90°$

(18) $\sec 0°$

(19) $\tan 0°$

(20) $\tan 30°$

(21) $\csc(\pi/2)$

(22) $\cot(\pi/4)$

(23) $\cos 30°$

(24) $\cot 30°$

Instructions: Determine the answer to each problem. Check your answers in the back of the book.

(1) csc 0°

(2) cot 45°

(3) tan 0°

(4) cot(π/2)

(5) sec 90°

(6) cos 0°

(7) cot 0°

(8) sec(π/2)

(9) cot 45°

(10) tan(π/2)

(11) sec(π/6)

(12) sin 30°

(13) sec 90°

(14) cot(π/6)

(15) csc(π/3)

(16) sin(π/2)

(17) tan(π/2)

(18) sec 30°

(19) sec 0°

(20) cos 90°

(21) sec 60°

(22) csc 90°

(23) cos(π/6)

(24) cos(π/2)

Instructions: Determine the answer to each problem. Check your answers in the back of the book.

(1) csc 90°

(2) tan 60°

(3) sin 0°

(4) cos 60°

(5) cos(π/3)

(6) csc(π/4)

(7) tan(π/6)

(8) csc 45°

(9) cot(π/3)

(10) cos(π/4)

(11) csc 90°

(12) sec(π/4)

(13) cot 30°

(14) cot(π/6)

(15) tan(π/3)

(16) cos 60°

(17) sec 90°

(18) csc(π/6)

(19) csc(π/3)

(20) sec(π/2)

(21) cos(π/2)

(22) cot(π/2)

(23) cos 90°

(24) sin 60°

Instructions: Determine the answer to each problem. Check your answers in the back of the book.

(1) sec 30°

(2) sin 45°

(3) csc 0°

(4) sin(π/2)

(5) cos(π/4)

(6) sin 45°

(7) csc 45°

(8) cos 45°

(9) cot(π/4)

(10) sec 0°

(11) cot 60°

(12) cos 0°

(13) tan 0°

(14) csc(π/3)

(15) csc 30°

(16) tan(π/4)

(17) sec(π/2)

(18) cot(π/2)

(19) cot 0°

(20) csc(π/4)

(21) tan(π/4)

(22) csc 45°

(23) csc 0°

(24) sec(π/3)

Instructions: Determine the answer to each problem. Check your answers in the back of the book.

(1) $\csc(\pi/6)$

(2) $\sec 90°$

(3) $\cos(\pi/4)$

(4) $\csc 0°$

(5) $\csc 45°$

(6) $\cot(\pi/4)$

(7) $\cos 30°$

(8) $\cot(\pi/3)$

(9) $\cos(\pi/3)$

(10) $\tan(\pi/2)$

(11) $\cos 0°$

(12) $\csc 60°$

(13) $\cos 0°$

(14) $\sin 30°$

(15) $\tan(\pi/3)$

(16) $\csc(\pi/6)$

(17) $\cot 30°$

(18) $\sec(\pi/4)$

(19) $\tan 90°$

(20) $\csc(\pi/2)$

(21) $\sec 0°$

(22) $\cos 45°$

(23) $\cos 90°$

(24) $\cos(\pi/2)$

Instructions: Determine the answer to each problem. Check your answers in the back of the book.

(1) sin(π/6)

(2) sin 60°

(3) cot 30°

(4) csc 0°

(5) csc(π/2)

(6) sec(π/2)

(7) tan(π/2)

(8) csc(π/4)

(9) cos(π/3)

(10) csc 0°

(11) sec(π/4)

(12) cot(π/4)

(13) sec 90°

(14) csc(π/2)

(15) sec 45°

(16) tan(π/4)

(17) cot 0°

(18) csc 30°

(19) cot(π/6)

(20) tan(π/6)

(21) cos 90°

(22) cot(π/4)

(23) csc 60°

(24) cos 30°

Instructions: Determine the answer to each problem. Check your answers in the back of the book.

(1) $\cot(\pi/3)$

(2) $\cot 60°$

(3) $\sec 0°$

(4) $\sin(\pi/6)$

(5) $\cos 0°$

(6) $\cos 90°$

(7) $\sec 0°$

(8) $\sin 0°$

(9) $\tan 0°$

(10) $\sin(\pi/6)$

(11) $\csc(\pi/6)$

(12) $\csc 0°$

(13) $\cot 45°$

(14) $\cos(\pi/6)$

(15) $\tan 90°$

(16) $\cos 90°$

(17) $\tan(\pi/4)$

(18) $\cot(\pi/6)$

(19) $\cos 0°$

(20) $\tan(\pi/6)$

(21) $\sec(\pi/4)$

(22) $\sin 90°$

(23) $\tan 90°$

(24) $\csc 90°$

Instructions: Determine the answer to each problem. Check your answers in the back of the book.

(1) sin(π/4)

(2) sin(π/6)

(3) csc 0°

(4) sin 30°

(5) sec(π/2)

(6) csc 90°

(7) tan 30°

(8) cos 60°

(9) tan(π/6)

(10) csc 0°

(11) tan 90°

(12) cos 0°

(13) sec(π/2)

(14) sin(π/2)

(15) sin 60°

(16) tan 60°

(17) tan 45°

(18) cos(π/6)

(19) csc 90°

(20) tan(π/6)

(21) tan 30°

(22) sin 45°

(23) tan(π/4)

(24) sin 0°

Instructions: Determine the answer to each problem. Check your answers in the back of the book.

(1) $\csc(\pi/6)$

(2) $\csc 30°$

(3) $\csc(\pi/6)$

(4) $\csc 0°$

(5) $\sin 90°$

(6) $\sec(\pi/3)$

(7) $\sin(\pi/4)$

(8) $\sec 0°$

(9) $\csc(\pi/3)$

(10) $\sin(\pi/6)$

(11) $\sec 45°$

(12) $\csc(\pi/6)$

(13) $\tan 0°$

(14) $\cos(\pi/2)$

(15) $\csc(\pi/4)$

(16) $\csc(\pi/6)$

(17) $\sec(\pi/6)$

(18) $\sin 60°$

(19) $\sin 45°$

(20) $\csc 60°$

(21) $\sec 30°$

(22) $\cot 90°$

(23) $\tan(\pi/3)$

(24) $\cos(\pi/6)$

Chapter 6: Finding the Reference Angle

If you need to evaluate a trig function at an angle between 90° and 360°, it is useful to first determine the reference angle. The reference angle is an angle between 0° and 90°, which gives the same value for the trig function apart from a possible minus sign. In this chapter, we will focus on how to determine the reference angle, and in Chapter 7 we will see how this relates to evaluating a trig function at an angle between 90° and 360°.

The standard convention for measuring angles in trigonometry is counterclockwise from the $+x$-axis, as illustrated below. With this system, 0° corresponds to the $+x$-axis, 90° corresponds to the $+y$-axis, 180° corresponds to the $-x$-axis, 270° corresponds to the $-y$-axis, and 360° is back to the $+x$-axis.

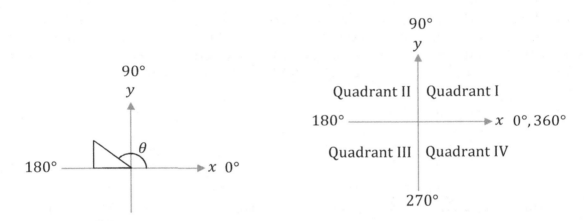

Angles between 0° and 360° can be divided into four quadrants. The standard convention for numbering the four quadrants is to use Roman numerals I, II, III, and IV as illustrated above. An angle between 0° and 90° is in Quadrant I, an angle between 90° and 180° is in Quadrant II, an angle between 180° and 270° is in Quadrant III, and an angle between 270° and 360° is in Quadrant IV.

Note that the angle is unchanged, geometrically, if you add 360° to it. For example, 45°, 405°, 765°, and −315° are all the same angle. Therefore, an angle between −90° and 0° lies in Quadrant IV and an angle between −180° and −90° lies in Quadrant III.

In trigonometry, it is often useful to refer to the unit circle. The unit circle lies in the xy plane, centered about the origin, and has a radius of 1 unit, as illustrated on the following page. For any angle between 0° and 360°, we can draw a corresponding right triangle in the unit circle with a hypotenuse of one unit and legs that are parallel to the x- and y-axes, also as illustrated on the following page. In Chapter 7, we will see how such a unit circle can help us determine the sign of the basic trig functions in each quadrant. For now, we will focus on how drawing a right triangle in the unit circle helps us find the reference angle.

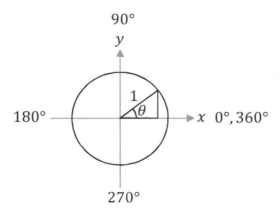

Consider the right triangle formed in the unit circle by the Quadrant II angle illustrated below. The reference angle is the corresponding Quadrant I angle for which the opposite and adjacent sides have the same lengths as they have for the original angle. In Quadrant II, the reference angle can be found by subtracting the Quadrant II angle from $180°$. For example, the reference angle for $120°$ is $180° - 120° = 60°$.

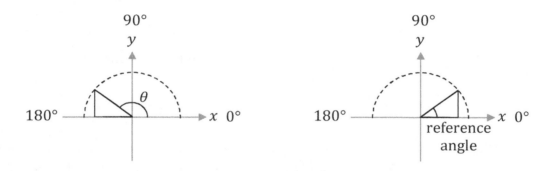

Consider the right triangle formed in the unit circle by the Quadrant III angle illustrated on the following page. In Quadrant III, the reference angle can be found by subtracting $180°$ from the Quadrant III angle. For example, the reference angle for $210°$ is $210° - 180° = 30°$. Now consider the right triangle formed in the unit circle by the Quadrant IV angle illustrated on the following page. In Quadrant IV, the reference angle can be found by subtracting the Quadrant IV angle from $360°$. For example, the reference angle for $345°$ is $360° - 345° = 15°$.

The reference angle is defined as the Quadrant I angle for which the right triangle formed in the unit circle has opposite and adjacent sides of the same length as the right triangle formed in the unit circle for the original Quadrant II, III, or IV angle. In this way, the sine, cosine, and tangent will have the same ratio for both right triangles (but may not have the same sign, as we will see in Chapter 7). Therefore, as we will explore in Chapter 7, the reference angle serves to help us evaluate the basic trig functions at angles in Quadrants II, III, and IV.

Take some time to understand these concepts, and study the examples. Once you understand the following examples, you are ready to practice the technique yourself. You may need to refer to the examples frequently as you begin, but should try to solve the exercises all by yourself once you get the hang of it. Be sure to check the answers at the back of the book to ensure that you are solving the problems correctly.

Instructions: Determine the reference angle for the given Quadrant II, III, or IV angle. Also, draw a unit circle, draw the right triangle formed in the unit circle for the given angle, draw the right triangle formed in the unit circle for the reference angle, label the given angle, and label the reference angle.

Procedure: If the given angle is negative, first add 360° to it. Next, determine in which quadrant the given angle lies.
- An angle between 90° and 180° is in Quadrant II.
- An angle between 180° and 270° is in Quadrant III.
- An angle between 270° and 360° is in Quadrant IV.

Finally, use the correct formula to find the reference angle:
- If the given angle lies in Quadrant II, subtract the given angle from 180°.
- If the given angle lies in Quadrant III, subtract 180° from the given angle.
- If the given angle lies in Quadrant IV, subtract the given angle from 360°.

Note: The examples that follow and the answers in the back of the book just give the reference angles (and not the diagrams). However, it is highly recommended that you draw the diagrams because this will help you to visualize the concepts and to understand the concepts geometrically.

Example 1: 100°. This angle is in Quadrant II: The reference angle is 180° − 100° = 80°.

Example 2: 290°. This angle is in Quadrant IV: The reference angle is 360° − 290° = 70°.

Example 3: −155°. First, add 360°: −155° + 360° = 205°. This angle is in Quadrant III: The reference angle is 205° − 180° = 25°.

81

Instructions: Determine the reference angle for the given Quadrant II, III, or IV angle. Also, draw and label a diagram. Check your answers in the back of the book.

(1) 203°

(2) 331°

(3) 188°

(4) 195°

(5) 341°

(6) −98°

(7) 109°

(8) 208°

(9) 315°

(10) 116°

(11) 220°

(12) −179°

(13) −104°

(14) 249°

(15) 135°

(16) 111°

(17) 288°

(18) −28°

Instructions: Determine the reference angle for the given Quadrant II, III, or IV angle. Also, draw and label a diagram. Check your answers in the back of the book.

(1) 336°

(2) −11°

(3) −160°

(4) 327°

(5) 211°

(6) 92°

(7) −145°

(8) −15°

(9) 230°

(10) 109°

(11) −36°

(12) 346°

(13) −79°

(14) 96°

(15) 297°

(16) 250°

(17) 176°

(18) 232°

Instructions: Determine the reference angle for the given Quadrant II, III, or IV angle. Also, draw and label a diagram. Check your answers in the back of the book.

(1) −49°   (2) 338°   (3) 200°

(4) 98°   (5) 312°   (6) 145°

(7) 99°   (8) 195°   (9) −120°

(10) 311°   (11) −52°   (12) 209°

(13) 140°   (14) 214°   (15) 332°

(16) 330°   (17) 98°   (18) 165°

Instructions: Determine the reference angle for the given Quadrant II, III, or IV angle. Also, draw and label a diagram. Check your answers in the back of the book.

(1) 227°

(2) 302°

(3) 161°

(4) −7°

(5) 159°

(6) 230°

(7) −163°

(8) 228°

(9) 136°

(10) 176°

(11) 154°

(12) 103°

(13) 220°

(14) −146°

(15) 195°

(16) 231°

(17) 345°

(18) 120°

Instructions: Determine the reference angle for the given Quadrant II, III, or IV angle. Also, draw and label a diagram. Check your answers in the back of the book.

(1) 253°

(2) 160°

(3) 304°

(4) 120°

(5) 196°

(6) 208°

(7) −45°

(8) 264°

(9) −4°

(10) 309°

(11) 116°

(12) 145°

(13) 322°

(14) 175°

(15) 140°

(16) −13°

(17) 122°

(18) 214°

Instructions: Determine the reference angle for the given Quadrant II, III, or IV angle. Also, draw and label a diagram. Check your answers in the back of the book.

(1) 169°

(2) 128°

(3) 300°

(4) 114°

(5) 142°

(6) 178°

(7) 290°

(8) −140°

(9) −25°

(10) 189°

(11) 116°

(12) 164°

(13) 284°

(14) −169°

(15) 157°

(16) 305°

(17) 240°

(18) −92°

Instructions: Determine the reference angle for the given Quadrant II, III, or IV angle. Also, draw and label a diagram. Check your answers in the back of the book.

(1) 209°

(2) 142°

(3) 306°

(4) −29°

(5) −148°

(6) 230°

(7) 210°

(8) 146°

(9) 353°

(10) 259°

(11) 94°

(12) −85°

(13) 297°

(14) −3°

(15) 359°

(16) 185°

(17) 228°

(18) −137°

Instructions: Determine the reference angle for the given Quadrant II, III, or IV angle. Also, draw and label a diagram. Check your answers in the back of the book.

(1) 283°   (2) −153°   (3) 202°

(4) 314°   (5) −58°   (6) 157°

(7) 153°   (8) 117°   (9) 108°

(10) 345°   (11) −11°   (12) 334°

(13) 241°   (14) 123°   (15) 195°

(16) 209°   (17) −84°   (18) 335°

Instructions: Determine the reference angle for the given Quadrant II, III, or IV angle. Also, draw and label a diagram. Check your answers in the back of the book.

(1) 220°

(2) −108°

(3) 151°

(4) 163°

(5) 284°

(6) −51°

(7) 209°

(8) 302°

(9) 121°

(10) 130°

(11) 355°

(12) 230°

(13) 347°

(14) 230°

(15) 315°

(16) −96°

(17) 283°

(18) 154°

Instructions: Determine the reference angle for the given Quadrant II, III, or IV angle. Also, draw and label a diagram. Check your answers in the back of the book.

(1) 345°

(2) 138°

(3) 150°

(4) 179°

(5) 349°

(6) 131°

(7) 121°

(8) 151°

(9) 131°

(10) 118°

(11) 127°

(12) 336°

(13) 183°

(14) −10°

(15) 215°

(16) 263°

(17) −94°

(18) 359°

Instructions: Determine the reference angle for the given Quadrant II, III, or IV angle. Also, draw and label a diagram. Check your answers in the back of the book.

(1) 321°   (2) 279°   (3) 311°

(4) −2°   (5) 97°   (6) 103°

(7) 251°   (8) 205°   (9) 130°

(10) 265°   (11) −113°   (12) 344°

(13) 251°   (14) 337°   (15) 254°

(16) −47°   (17) 298°   (18) 153°

Instructions: Determine the reference angle for the given Quadrant II, III, or IV angle. Also, draw and label a diagram. Check your answers in the back of the book.

(1) 239°

(2) 325°

(3) 163°

(4) 145°

(5) 222°

(6) −162°

(7) 330°

(8) −55°

(9) 91°

(10) 257°

(11) 306°

(12) 126°

(13) 146°

(14) 296°

(15) 102°

(16) 111°

(17) 308°

(18) 109°

Chapter 7: Determine Basic Trig Functions in Quadrants II-IV

Consider the right triangle formed in the unit circle, which we described in the introduction to Chapter 6, and which is illustrated in Quadrant I below. The hypotenuse of the right triangle equals the radius of the circle, and since it is a unit circle, the radius is one. The sine function equals the opposite over the hypotenuse, while the cosine function equals the adjacent over the hypotenuse. Since the hypotenuse is one, for the right triangle formed in the unit circle, the sine function equals the opposite side and the cosine function equals the adjacent side. The adjacent side is x, while the opposite side is y. Therefore, for the right triangle formed in the unit circle, $x = \cos\theta$ and $y = \sin\theta$.

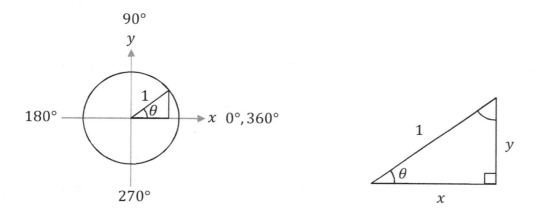

Recall how we defined the reference angle in Chapter 6: The reference angle is the Quadrant I angle for which the right triangle formed in the unit circle has opposite and adjacent sides of the same length as the right triangle formed in the unit circle for the original Quadrant II, III, or IV angle. This means that x and y have the same length, but possibly different sign, in the original right triangle as they do for the Quadrant I right triangle involving the reference angle. Since $x = \cos\theta$ and $y = \sin\theta$, this means that all of the trig functions evaluated in Quadrants II, III, and IV are the same as the same trig function evaluated at the reference angle – apart from a possible minus sign.

Therefore, if you want to evaluate a trig function in Quadrants II, III, or IV, you need do two things: First, determine the reference angle; and, next, determine if the answer is positive or negative. For example, if you want to find $\cos 150°$, first you need to determine that the reference angle is $180° - 150° = 30°$ (see Chapter 6), and then you need to determine if $\cos 150°$ is positive or negative. You should already know that $\cos 30° = \sqrt{3}/2$ (see Chapter 5). Momentarily, we will learn that the cosine function is negative in Quadrant II, which means that $\cos 150° = -\sqrt{3}/2$.

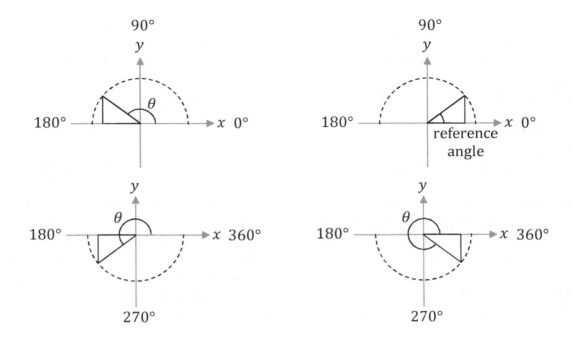

Recall that $\sin\theta$ equals y for the right triangle formed in the unit circle. This means that the sine function has the same sign as y: $\sin\theta$ is positive in Quadrants I and II, and it is negative in Quadrants III and IV. Similarly, $\cos\theta$ equals x for the right triangle formed in the unit circle: $\cos\theta$ is positive in Quadrants I and IV, and negative in Quadrants II and III.

Recall that $\tan\theta$ equals $\sin\theta / \cos\theta$. In Quadrant II, $\tan\theta$ is negative because $\sin\theta$ is positive while $\cos\theta$ is negative. In Quadrant III, $\tan\theta$ is positive because both $\sin\theta$ and $\cos\theta$ are negative. In Quadrant IV, $\tan\theta$ is negative because $\sin\theta$ is negative while $\cos\theta$ is positive.

Finally, $\csc\theta$, $\sec\theta$, and $\cot\theta$ have the same signs as $\sin\theta$, $\cos\theta$, and $\tan\theta$, respectively. The signs of the various trig functions in each Quadrant are tabulated below.

Quadrant	$\sin\theta$	$\cos\theta$	$\tan\theta$	$\sec\theta$	$\csc\theta$	$\cot\theta$
I	+	+	+	+	+	+
II	+	−	−	−	+	−
III	−	−	+	−	−	+
IV	−	+	−	+	−	−

Take some time to understand these concepts, and study the examples. Once you understand the following examples, you are ready to practice the technique yourself. You may need to refer to the examples frequently as you begin, but should try to solve the exercises all by yourself once you get the hang of it. Be sure to check the answers at the back of the book to ensure that you are solving the problems correctly.

Instructions: Determine the answer to each problem. Some of the angles are given in degrees, while some angles are given in radians. Degrees will be indicated with a degree symbol (°), whereas angles in radians will have a π, but the radians will be implied.

Procedure: First, determine the reference angle using the technique described in Chapter 6. Evaluate the trig function at the reference angle using the strategy from Chapter 5. Next, determine which Quadrant the given angle lies in, and whether the indicated trig function is positive or negative in that Quadrant (see the table on the previous page). If the trig function is negative in that Quadrant, include a minus sign. If the angle is given in radians instead of degrees, you can convert the angle to degrees using the method of Chapter 2.

Rationalize the denominator. For example, $\frac{1}{\sqrt{2}} = \frac{1}{\sqrt{2}}\frac{\sqrt{2}}{\sqrt{2}} = \frac{\sqrt{2}}{2}$ and $\frac{1}{\sqrt{3}} = \frac{1}{\sqrt{3}}\frac{\sqrt{3}}{\sqrt{3}} = \frac{\sqrt{3}}{3}$.

Example 1: $\sin 120°$. The reference angle is $180° - 120° = 60°$. You should know that $\sin 60° = \sqrt{3}/2$. The given angle, 120°, lies in Quadrant II; the sine function is positive in Quadrant II. Therefore, $\sin 120° = \sin 60° = \sqrt{3}/2$.

Example 2: $\sec 210°$. The reference angle is $210° - 180° = 30°$. You should know that $\sec 30° = \frac{1}{\cos 30°} = \frac{1}{\sqrt{3}/2} = \frac{2}{\sqrt{3}} = \frac{2}{\sqrt{3}}\frac{\sqrt{3}}{\sqrt{3}} = \frac{2\sqrt{3}}{3}$. The given angle, 210°, lies in Quadrant III; the secant function is negative in Quadrant III (since cosine is negative in Quadrant III). Therefore, $\sec 210° = -\sec 30° = -2\sqrt{3}/3$.

Example 3: $\tan 180°$. The reference angle is $180° - 180° = 0°$. You should know that $\tan 0° = \frac{\sin 0°}{\cos 0°} = \frac{0}{1} = 0$. The answer is 0, so the sign doesn't matter: $\tan 180° = \tan 0° = 0$.

Example 4: $\csc 270°$. The reference angle is $360° - 270° = 90°$. You should know that $\csc 90° = \frac{1}{\sin 90°} = \frac{1}{1} = 1$. The given angle, 270°, lies on the border of Quadrants III and IV; the cosecant function is negative in both of these quadrants (since sine is negative in Quadrants III and IV). Therefore, $\csc 270° = -\csc 90° = -1$.

Example 5: $\cot 330°$. The reference angle is $360° - 330° = 30°$. You should know that $\cot 30° = \frac{1}{\tan 30°} = \frac{\cos 30°}{\sin 30°} = \frac{\sqrt{3}/2}{1/2} = \frac{\sqrt{3}}{2}\frac{2}{1} = \sqrt{3}$. To divide two fractions, multiply the first fraction by the reciprocal of the second fraction. The given angle, 330°, lies in Quadrant IV; the cotangent function is negative in Quadrant IV. Therefore, $\cot 330° = -\cot 30° = -\sqrt{3}$.

Instructions: Determine the answer to each problem. Check your answers in the back of the book.

(1) csc 150°

(2) sec(5π/4)

(3) cot 315°

(4) tan 135°

(5) sin 330°

(6) csc(5π/3)

(7) cos 225°

(8) sin(5π/6)

(9) cos 210°

(10) csc(3π/4)

(11) cot(2π/3)

(12) sec(4π/3)

(13) cot 150°

(14) sec(2π)

(15) cos(5π/6)

(16) csc(3π/2)

(17) csc 180°

(18) csc 225°

(19) csc 240°

(20) csc 225°

(21) cos(π)

(22) sec 180°

(23) tan(5π/3)

(24) tan(5π/4)

Instructions: Determine the answer to each problem. Check your answers in the back of the book.

(1) $\cot(11\pi/6)$

(2) $\sin 270°$

(3) $\sec 180°$

(4) $\sec 360°$

(5) $\csc 120°$

(6) $\cos(3\pi/4)$

(7) $\cos(2\pi/3)$

(8) $\cos 300°$

(9) $\cot(5\pi/3)$

(10) $\cot 315°$

(11) $\cot(\pi)$

(12) $\sec(2\pi)$

(13) $\sin 300°$

(14) $\sec(7\pi/6)$

(15) $\csc 210°$

(16) $\csc(3\pi/2)$

(17) $\sec(5\pi/6)$

(18) $\sin(5\pi/3)$

(19) $\sec 270°$

(20) $\sec 150°$

(21) $\sec 300°$

(22) $\cos(3\pi/2)$

(23) $\cot 315°$

(24) $\tan(7\pi/4)$

Instructions: Determine the answer to each problem. Check your answers in the back of the book.

(1) tan 270°

(2) cot(2π/3)

(3) sin 150°

(4) cot(3π/4)

(5) csc 225°

(6) csc(5π/3)

(7) sin(4π/3)

(8) cos 330°

(9) sin(5π/6)

(10) cot 300°

(11) sec 120°

(12) tan 150°

(13) sin(11π/6)

(14) cos(7π/6)

(15) sec 135°

(16) tan 120°

(17) sin 330°

(18) cot(π)

(19) cot 360°

(20) sin 240°

(21) cos(7π/4)

(22) csc(5π/4)

(23) cot 135°

(24) sin(4π/3)

Instructions: Determine the answer to each problem. Check your answers in the back of the book.

(1) cot 180°

(2) sin(π)

(3) tan(5π/6)

(4) csc(7π/6)

(5) tan 210°

(6) tan 330°

(7) csc 210°

(8) sec 270°

(9) sec(2π)

(10) csc 360°

(11) sec(5π/4)

(12) sin 150°

(13) csc(5π/6)

(14) cos(π)

(15) sec (2π/3)

(16) sin(11π/6)

(17) csc(3π/4)

(18) sin 180°

(19) cot(3π/2)

(20) cot(π)

(21) tan 135°

(22) cos 225°

(23) cos(5π/3)

(24) tan(7π/4)

Instructions: Determine the answer to each problem. Check your answers in the back of the book.

(1) cot(5π/4)

(2) sin(5π/3)

(3) tan 150°

(4) cos 360°

(5) tan 240°

(6) sin 180°

(7) cos 315°

(8) cos(7π/4)

(9) cot 300°

(10) sin(π)

(11) csc 210°

(12) tan 225°

(13) cos 360°

(14) sin 135°

(15) cos 330°

(16) csc(7π/4)

(17) cos(2π)

(18) sec 180°

(19) sin(5π/6)

(20) csc(3π/4)

(21) csc 150°

(22) csc(7π/4)

(23) tan(5π/4)

(24) sec(7π/6)

Instructions: Determine the answer to each problem. Check your answers in the back of the book.

(1) cot 300°

(2) tan 315°

(3) sec 135°

(4) sec 360°

(5) tan 330°

(6) $\cos(3\pi/2)$

(7) $\csc(5\pi/6)$

(8) cos 135°

(9) sin 210°

(10) tan 240°

(11) $\cos(2\pi)$

(12) sec 360°

(13) $\sec(5\pi/3)$

(14) $\cot(5\pi/4)$

(15) cos 150°

(16) sin 300°

(17) cos 120°

(18) sec 330°

(19) csc 270°

(20) $\tan(3\pi/4)$

(21) $\csc(5\pi/3)$

(22) $\sin(3\pi/2)$

(23) sec 210°

(24) $\cos(2\pi/3)$

Instructions: Determine the answer to each problem. Check your answers in the back of the book.

(1) sec 330°

(2) csc(7π/6)

(3) cos 180°

(4) sin(5π/3)

(5) cos 210°

(6) cot 360°

(7) sec(2π)

(8) sec 360°

(9) cot 135°

(10) cot 120°

(11) sec 300°

(12) cot(π)

(13) sec 225°

(14) cos(11π/6)

(15) cot 135°

(16) sin 150°

(17) sin 330°

(18) tan(2π/3)

(19) cot(5π/4)

(20) sec 120°

(21) tan(7π/6)

(22) cos(7π/4)

(23) cos 300°

(24) cot 150°

Instructions: Determine the answer to each problem. Check your answers in the back of the book.

(1) sec 150°

(2) cot(7π/6)

(3) csc 225°

(4) cos 210°

(5) csc(3π/4)

(6) cot(5π/6)

(7) sec(2π)

(8) cot 180°

(9) cos 360°

(10) sec 180°

(11) tan(π)

(12) sin 225°

(13) sin(7π/6)

(14) cot 330°

(15) cot(5π/6)

(16) sin 150°

(17) tan(5π/3)

(18) sec 240°

(19) sec 180°

(20) csc(7π/6)

(21) csc 210°

(22) tan(7π/6)

(23) csc(5π/4)

(24) cot 180°

Instructions: Determine the answer to each problem. Check your answers in the back of the book.

(1) $\cot(2\pi)$

(2) $\cot 240°$

(3) $\csc 360°$

(4) $\tan(11\pi/6)$

(5) $\sec 330°$

(6) $\cos(11\pi/6)$

(7) $\sin 225°$

(8) $\tan 315°$

(9) $\cos 180°$

(10) $\cos(\pi)$

(11) $\sin(7\pi/4)$

(12) $\sin(7\pi/6)$

(13) $\tan(7\pi/6)$

(14) $\cos 210°$

(15) $\sec 120°$

(16) $\cos 150°$

(17) $\cos(7\pi/4)$

(18) $\sin(5\pi/4)$

(19) $\csc(2\pi)$

(20) $\cos 330°$

(21) $\cos 270°$

(22) $\csc 360°$

(23) $\tan(3\pi/2)$

(24) $\tan 225°$

Instructions: Determine the answer to each problem. Check your answers in the back of the book.

(1) tan 180°

(2) tan 210°

(3) tan(π)

(4) sec(2π/3)

(5) sin(7π/4)

(6) csc 360°

(7) csc(4π/3)

(8) csc(π)

(9) sec 180°

(10) sec 240°

(11) cos 330°

(12) tan 135°

(13) tan(5π/3)

(14) cot(π)

(15) tan(11π/6)

(16) sin(3π/2)

(17) cos 330°

(18) sec 360°

(19) tan 135°

(20) sin 240°

(21) csc 360°

(22) sec 225°

(23) csc 315°

(24) sec(4π/3)

Instructions: Determine the answer to each problem. Check your answers in the back of the book.

(1) cot(5π/4)

(2) sin 120°

(3) sec 270°

(4) sin(5π/6)

(5) tan(2π/3)

(6) cot 210°

(7) cot 150°

(8) csc(7π/6)

(9) csc(5π/6)

(10) sin 210°

(11) sec 360°

(12) csc 315°

(13) tan(7π/4)

(14) cot 225°

(15) cot 300°

(16) cot 315°

(17) csc(5π/4)

(18) sin(7π/6)

(19) sin(2π)

(20) cot(3π/2)

(21) cos 270°

(22) cot 150°

(23) cos 240°

(24) tan(5π/3)

Instructions: Determine the answer to each problem. Check your answers in the back of the book.

(1) sec 135°

(2) csc(3π/2)

(3) cot(11π/6)

(4) cot(3π/4)

(5) csc 360°

(6) cos 270°

(7) csc(3π/2)

(8) cos 330°

(9) tan 210°

(10) cot(4π/3)

(11) sec 315°

(12) sin(11π/6)

(13) cot 180°

(14) tan 120°

(15) cos(2π)

(16) cos(2π/3)

(17) cos 360°

(18) cot(2π/3)

(19) sec 330°

(20) sin 270°

(21) sec 315°

(22) sin 150°

(23) csc(5π/6)

(24) cos(7π/4)

Chapter 8: The Inverse Trig Functions

The function $\sin 30°$ equals one-half because the side opposite to $30°$ is always one-half as long as the hypotenuse in a $30°$-$60°$-$90°$ triangle. The sine function tells us what the ratio of the opposite side to the hypotenuse is for a given angle.

There also exists an inverse sine function ($\sin^{-1}$), which tells us what angle makes the opposite side over the hypotenuse equal to a specified ratio. For example, $\sin^{-1}\left(\frac{1}{2}\right)$ equals $30°$ or $150°$. The inverse sine of one-half, which we express as $\sin^{-1}\left(\frac{1}{2}\right)$, asks what angle you could take the sine of and get an answer of one-half. (It is important to note that $\sin^{-1}$ does **not** mean to find one over the sine function – it is not a multiplicative inverse.)

When we use the sine function, the argument is an angle and the answer is a fraction between -1 and $+1$. When we use the inverse sine function, the argument is a fraction between -1 and $+1$ and the answer is an angle.

Consider the equation $x = \sin\theta$. If you want to solve for θ in this equation, take the inverse sine of both sides: $\sin^{-1}x = \sin^{-1}(\sin\theta) = \theta$, or $\theta = \sin^{-1}x$. Note that $\sin^{-1}(\sin\theta) = \theta$. That is, if you take the sine of angle and then take the inverse sine of the resulting fraction, you get back the original angle as a solution.

If you know how to find the sine of an angle, it's easy to find the inverse sine of a fraction – it's essentially the same thing backwards. For example, since $\sin 90° = 1$, it follows that $\sin^{-1}1 = 90°$. The other inverse trig functions work the same way. For example, since $\tan 120°$ and $\tan 300°$ both equal $-\sqrt{3}$, it follows that $\tan^{-1}(-\sqrt{3})$ equals $120°$ or $300°$.

In general there are two possible answers to an inverse trig function because each trig function is positive and negative in two different quadrants. For example, $\cos^{-1}\left(-\frac{1}{2}\right)$ equals $120°$ or $240°$ because $\cos 120° = \cos 240° = -\frac{1}{2}$.

Take some time to understand these concepts, and study the examples. Once you understand the following examples, you are ready to practice the technique yourself. You may need to refer to the examples frequently as you begin, but should try to solve the exercises all by yourself once you get the hang of it. Be sure to check the answers at the back of the book to ensure that you are solving the problems correctly.

Instructions: Determine the two angles that solve each problem. Check your answers by applying the trig function to each angle and see if you obtain the same ratio – including the same sign – as the given argument of the inverse trig function.

Procedure: To find the inverse trig function of a fraction, first find the reference angle – i.e. which Quadrant I angle could you apply the trig function to in order find the absolute value of the fraction? Then determine the two angles in Quadrants I-IV corresponding to the reference angle that give the specified trig function the correct sign.

If the problem gives you an inverse secant, inverse cosecant, or inverse cotangent, first find the reciprocal of the argument and then treat it as an inverse cosine, sine, or tangent, respectively, because $\sec\theta = \frac{1}{\cos\theta}$, $\csc\theta = \frac{1}{\sin\theta}$, and $\cot\theta = \frac{1}{\tan\theta}$. For example, $\sec^{-1}(2)$ is the same as $\cos^{-1}\left(\frac{1}{2}\right)$, which is $60°$ or $300°$.

There may only be one answer if the answer lies on the x- or y-axis – i.e. if the answer is $0°, 90°, 180°,$ or $270°$.

Example 1: $\cos^{-1}\left(\frac{\sqrt{3}}{2}\right)$. The Quadrant I reference angle for which the cosine of the angle would equal $\frac{\sqrt{3}}{2}$ is $30°$: That is, $\cos 30° = \frac{\sqrt{3}}{2}$. The cosine function is positive in Quadrants I and IV. The Quadrant IV angle corresponding to $30°$ is $360° - 30° = 330°$. The answers to this example are $30°$ and $330°$.

Example 2: $\csc^{-1}(-1)$. The Quadrant I reference angle for which the cosecant of the angle would equal 1 is the same as the Quadrant I reference angle for which the sine of the angle would equal 1 (since $\csc\theta = \frac{1}{\sin\theta}$ and $\frac{1}{1} = 1$). This is $90°$: That is, $\sin 90° = 1$ and $\csc 90° = 1$. But we need -1, not $+1$. Since $\sin 270° = -1$ and $\csc 270° = -1$, there is just one answer to this problem, $270°$ (which lies on the border of Quadrants III and IV).

Example 3: $\tan^{-1}(-1)$. The Quadrant I reference angle for which the tangent of the angle would equal 1 is $45°$ (since $\tan 45° = \frac{\sin 45°}{\cos 45°}$ and $\sin 45° = \cos 45° = \frac{\sqrt{2}}{2}$): That is, $\tan 45° = 1$. The tangent function is negative in Quadrants II and IV. The Quadrant II and IV angles corresponding to $45°$ are $180° - 45° = 135°$ and $360° - 45° = 315°$. The answers to this example are $135°$ and $315°$.

Example 4: $\sec^{-1}(-2)$. The Quadrant I reference angle for which the secant of the angle would equal 2 is the same as the Quadrant I reference angle for which the cosine of the angle would equal $1/2$ (since $\sec\theta = \frac{1}{\cos\theta}$). This is $60°$: That is, $\cos 60° = 1/2$ and $\sec 60° = 2$. The secant and cosine functions are negative in Quadrants II and III. The Quadrant II and III angles corresponding to $60°$ are $180° - 60° = 120°$ and $180° + 60° = 240°$. The answers to this example are $120°$ and $240°$.

Example 5: $\cot^{-1}(0)$. The Quadrant I reference angle for which the cotangent of the angle would equal 0 is the same as the Quadrant I reference angle for which the tangent of the angle would be undefined (since $\cot\theta = \frac{1}{\tan\theta}$): This is $90°$: That is, $\tan 90°$ is undefined and $\cot 90° = 0$. The alternate angle is $270°$. The answers to this example are $90°$ and $270°$.

Instructions: Determine the answers to each problem. Most of the problems have two answers (in two different Quadrants). Check your answers in the back of the book.

(1) $\cos^{-1}\left(-\frac{1}{2}\right)$

(2) $\cot^{-1}(-\sqrt{3})$

(3) $\sin^{-1}(0)$

(4) $\sec^{-1}(2)$

(5) $\sec^{-1}(\sqrt{2})$

(6) $\sin^{-1}\left(-\frac{\sqrt{2}}{2}\right)$

(7) $\csc^{-1}(-2)$

(8) $\csc^{-1}(-\sqrt{2})$

(9) $\sec^{-1}\left(-\frac{2\sqrt{3}}{3}\right)$

(10) $\cot^{-1}(-1)$

(11) $\cos^{-1}\left(-\frac{1}{2}\right)$

(12) $\sec^{-1}(-2)$

(13) $\cos^{-1}\left(-\frac{\sqrt{3}}{2}\right)$

(14) $\sec^{-1}(2)$

(15) $\csc^{-1}(-2)$

(16) $\sec^{-1}(\sqrt{2})$

(17) $\cos^{-1}(0)$

(18) $\sin^{-1}\left(\frac{\sqrt{3}}{2}\right)$

(19) $\cos^{-1}(1)$

(20) $\sec^{-1}(1)$

(21) $\cot^{-1}(-\sqrt{3})$

(22) $\cot^{-1}(-1)$

(23) $\sin^{-1}\left(\frac{1}{2}\right)$

(24) $\cot^{-1}(\text{undefined})$

Instructions: Determine the answers to each problem. Most of the problems have two answers (in two different Quadrants). Check your answers in the back of the book.

(1) $\csc^{-1}(-2)$

(2) $\cos^{-1}\left(-\frac{1}{2}\right)$

(3) $\sec^{-1}(-2)$

(4) $\csc^{-1}(1)$

(5) $\sec^{-1}(-2)$

(6) $\csc^{-1}(2)$

(7) $\cot^{-1}\left(-\sqrt{3}\right)$

(8) $\tan^{-1}(0)$

(9) $\csc^{-1}\left(-\frac{2\sqrt{3}}{3}\right)$

(10) $\cos^{-1}(1)$

(11) $\csc^{-1}\left(\frac{2\sqrt{3}}{3}\right)$

(12) $\sin^{-1}\left(\frac{\sqrt{3}}{2}\right)$

(13) $\csc^{-1}(-2)$

(14) $\cot^{-1}\left(-\sqrt{3}\right)$

(15) $\cos^{-1}\left(\frac{\sqrt{2}}{2}\right)$

(16) $\sin^{-1}(0)$

(17) $\sec^{-1}\left(-\frac{2\sqrt{3}}{3}\right)$

(18) $\csc^{-1}\left(-\frac{2\sqrt{3}}{3}\right)$

(19) $\csc^{-1}\left(\sqrt{2}\right)$

(20) $\sin^{-1}\left(-\frac{\sqrt{2}}{2}\right)$

(21) $\cos^{-1}(0)$

(22) $\sec^{-1}(1)$

(23) $\cot^{-1}\left(\sqrt{3}\right)$

(24) $\sec^{-1}(2)$

Instructions: Determine the answers to each problem. Most of the problems have two answers (in two different Quadrants). Check your answers in the back of the book.

(1) $\sin^{-1}\left(\frac{\sqrt{3}}{2}\right)$

(2) $\cos^{-1}\left(\frac{1}{2}\right)$

(3) $\cot^{-1}(1)$

(4) $\csc^{-1}(\text{undefined})$

(5) $\tan^{-1}\left(-\frac{\sqrt{3}}{3}\right)$

(6) $\cot^{-1}(-1)$

(7) $\tan^{-1}(1)$

(8) $\cos^{-1}\left(-\frac{\sqrt{2}}{2}\right)$

(9) $\sec^{-1}(-\sqrt{2})$

(10) $\cot^{-1}(\text{undefined})$

(11) $\sin^{-1}\left(-\frac{\sqrt{2}}{2}\right)$

(12) $\tan^{-1}(-\sqrt{3})$

(13) $\cot^{-1}(-\sqrt{3})$

(14) $\tan^{-1}(1)$

(15) $\cos^{-1}(0)$

(16) $\sin^{-1}\left(-\frac{1}{2}\right)$

(17) $\sin^{-1}(1)$

(18) $\sec^{-1}(\sqrt{2})$

(19) $\cos^{-1}\left(\frac{1}{2}\right)$

(20) $\cos^{-1}\left(-\frac{\sqrt{2}}{2}\right)$

(21) $\csc^{-1}\left(-\frac{2\sqrt{3}}{3}\right)$

(22) $\cot^{-1}(\sqrt{3})$

(23) $\tan^{-1}\left(\frac{\sqrt{3}}{3}\right)$

(24) $\sec^{-1}(\text{undefined})$

Instructions: Determine the answers to each problem. Most of the problems have two answers (in two different Quadrants). Check your answers in the back of the book.

(1) $\sin^{-1}\left(\frac{1}{2}\right)$

(2) $\cot^{-1}(\text{undefined})$

(3) $\csc^{-1}(-1)$

(4) $\cos^{-1}\left(-\frac{\sqrt{2}}{2}\right)$

(5) $\tan^{-1}\left(-\frac{\sqrt{3}}{3}\right)$

(6) $\sin^{-1}(1)$

(7) $\sin^{-1}(0)$

(8) $\cot^{-1}\left(-\sqrt{3}\right)$

(9) $\csc^{-1}(\text{undefined})$

(10) $\cot^{-1}(\text{undefined})$

(11) $\tan^{-1}(0)$

(12) $\sec^{-1}(-2)$

(13) $\tan^{-1}(-1)$

(14) $\cos^{-1}\left(-\frac{\sqrt{3}}{2}\right)$

(15) $\sin^{-1}\left(\frac{\sqrt{2}}{2}\right)$

(16) $\cos^{-1}\left(\frac{1}{2}\right)$

(17) $\cot^{-1}(-1)$

(18) $\tan^{-1}(-1)$

(19) $\sin^{-1}\left(\frac{\sqrt{3}}{2}\right)$

(20) $\cot^{-1}\left(\frac{\sqrt{3}}{3}\right)$

(21) $\sec^{-1}(1)$

(22) $\cot^{-1}(0)$

(23) $\sin^{-1}(0)$

(24) $\cos^{-1}\left(\frac{1}{2}\right)$

Instructions: Determine the answers to each problem. Most of the problems have two answers (in two different Quadrants). Check your answers in the back of the book.

(1) $\cos^{-1}\left(\frac{\sqrt{2}}{2}\right)$

(2) $\sec^{-1}\left(-\sqrt{2}\right)$

(3) $\sec^{-1}(-2)$

(4) $\sin^{-1}\left(\frac{1}{2}\right)$

(5) $\sec^{-1}(1)$

(6) $\cos^{-1}(0)$

(7) $\tan^{-1}\left(-\frac{\sqrt{3}}{3}\right)$

(8) $\csc^{-1}(2)$

(9) $\sin^{-1}(0)$

(10) $\csc^{-1}(-1)$

(11) $\cos^{-1}(1)$

(12) $\csc^{-1}\left(\sqrt{2}\right)$

(13) $\csc^{-1}\left(-\frac{2\sqrt{3}}{3}\right)$

(14) $\sec^{-1}(\text{undefined})$

(15) $\csc^{-1}(-2)$

(16) $\sin^{-1}\left(\frac{\sqrt{2}}{2}\right)$

(17) $\sec^{-1}(1)$

(18) $\cot^{-1}(-1)$

(19) $\sec^{-1}\left(-\sqrt{2}\right)$

(20) $\cos^{-1}\left(-\frac{1}{2}\right)$

(21) $\cos^{-1}\left(\frac{\sqrt{3}}{2}\right)$

(22) $\csc^{-1}(-2)$

(23) $\sec^{-1}\left(-\frac{2\sqrt{3}}{3}\right)$

(24) $\csc^{-1}\left(\frac{2\sqrt{3}}{3}\right)$

Instructions: Determine the answers to each problem. Most of the problems have two answers (in two different Quadrants). Check your answers in the back of the book.

(1) $\cot^{-1}(-1)$

(2) $\tan^{-1}\left(\frac{\sqrt{3}}{3}\right)$

(3) $\cot^{-1}\left(\sqrt{3}\right)$

(4) $\cot^{-1}\left(-\sqrt{3}\right)$

(5) $\tan^{-1}(-1)$

(6) $\csc^{-1}(1)$

(7) $\tan^{-1}\left(-\frac{\sqrt{3}}{3}\right)$

(8) $\tan^{-1}\left(\sqrt{3}\right)$

(9) $\csc^{-1}(2)$

(10) $\sec^{-1}(-2)$

(11) $\sec^{-1}(2)$

(12) $\sin^{-1}\left(\frac{\sqrt{2}}{2}\right)$

(13) $\csc^{-1}\left(\frac{2\sqrt{3}}{3}\right)$

(14) $\tan^{-1}\left(-\sqrt{3}\right)$

(15) $\cot^{-1}(0)$

(16) $\sec^{-1}(\text{undefined})$

(17) $\sin^{-1}\left(\frac{\sqrt{3}}{2}\right)$

(18) $\csc^{-1}\left(\frac{2\sqrt{3}}{3}\right)$

(19) $\tan^{-1}(0)$

(20) $\cos^{-1}\left(-\frac{1}{2}\right)$

(21) $\cos^{-1}(0)$

(22) $\tan^{-1}\left(\frac{\sqrt{3}}{3}\right)$

(23) $\csc^{-1}(-2)$

(24) $\cos^{-1}\left(-\frac{\sqrt{2}}{2}\right)$

Instructions: Determine the answers to each problem. Most of the problems have two answers (in two different Quadrants). Check your answers in the back of the book.

(1) $\tan^{-1}\left(-\sqrt{3}\right)$

(2) $\csc^{-1}\left(-\frac{2\sqrt{3}}{3}\right)$

(3) $\csc^{-1}(-1)$

(4) $\sec^{-1}\left(\sqrt{2}\right)$

(5) $\sec^{-1}(2)$

(6) $\cos^{-1}\left(-\frac{\sqrt{3}}{2}\right)$

(7) $\cos^{-1}(1)$

(8) $\sec^{-1}\left(-\frac{2\sqrt{3}}{3}\right)$

(9) $\cot^{-1}\left(-\frac{\sqrt{3}}{3}\right)$

(10) $\csc^{-1}\left(-\sqrt{2}\right)$

(11) $\sec^{-1}(\text{undefined})$

(12) $\cot^{-1}\left(-\sqrt{3}\right)$

(13) $\cot^{-1}\left(\sqrt{3}\right)$

(14) $\csc^{-1}\left(\sqrt{2}\right)$

(15) $\tan^{-1}(-1)$

(16) $\cos^{-1}\left(\frac{\sqrt{2}}{2}\right)$

(17) $\sec^{-1}\left(\sqrt{2}\right)$

(18) $\cot^{-1}(\text{undefined})$

(19) $\cot^{-1}\left(-\sqrt{3}\right)$

(20) $\sec^{-1}(2)$

(21) $\csc^{-1}\left(-\frac{2\sqrt{3}}{3}\right)$

(22) $\cot^{-1}(1)$

(23) $\csc^{-1}(-1)$

(24) $\sin^{-1}\left(-\frac{1}{2}\right)$

Instructions: Determine the answers to each problem. Most of the problems have two answers (in two different Quadrants). Check your answers in the back of the book.

(1) $\csc^{-1}(1)$

(2) $\tan^{-1}(0)$

(3) $\csc^{-1}\left(-\frac{2\sqrt{3}}{3}\right)$

(4) $\cot^{-1}(1)$

(5) $\tan^{-1}\left(-\frac{\sqrt{3}}{3}\right)$

(6) $\tan^{-1}(-1)$

(7) $\cos^{-1}(1)$

(8) $\sin^{-1}\left(\frac{1}{2}\right)$

(9) $\sin^{-1}(0)$

(10) $\cot^{-1}(-1)$

(11) $\cos^{-1}(0)$

(12) $\sin^{-1}\left(\frac{1}{2}\right)$

(13) $\cot^{-1}(0)$

(14) $\cos^{-1}\left(-\frac{1}{2}\right)$

(15) $\sec^{-1}(-\sqrt{2})$

(16) $\cot^{-1}\left(-\frac{\sqrt{3}}{3}\right)$

(17) $\sin^{-1}\left(-\frac{\sqrt{2}}{2}\right)$

(18) $\csc^{-1}(1)$

(19) $\sec^{-1}(-2)$

(20) $\sin^{-1}(0)$

(21) $\cot^{-1}(\text{undefined})$

(22) $\sin^{-1}\left(-\frac{1}{2}\right)$

(23) $\csc^{-1}(-1)$

(24) $\tan^{-1}(-1)$

Instructions: Determine the answers to each problem. Most of the problems have two answers (in two different Quadrants). Check your answers in the back of the book.

(1) $\csc^{-1}(-1)$

(2) $\cos^{-1}\left(-\frac{\sqrt{3}}{2}\right)$

(3) $\cot^{-1}(-1)$

(4) $\cot^{-1}(\text{undefined})$

(5) $\sec^{-1}(1)$

(6) $\csc^{-1}(\text{undefined})$

(7) $\csc^{-1}(-1)$

(8) $\sin^{-1}\left(\frac{1}{2}\right)$

(9) $\tan^{-1}(-1)$

(10) $\tan^{-1}(0)$

(11) $\csc^{-1}(-2)$

(12) $\sec^{-1}(2)$

(13) $\sec^{-1}(-2)$

(14) $\cos^{-1}\left(\frac{1}{2}\right)$

(15) $\sin^{-1}(-1)$

(16) $\sin^{-1}\left(\frac{\sqrt{2}}{2}\right)$

(17) $\sec^{-1}(-1)$

(18) $\cos^{-1}\left(\frac{\sqrt{2}}{2}\right)$

(19) $\sin^{-1}\left(-\frac{\sqrt{2}}{2}\right)$

(20) $\tan^{-1}(\text{undefined})$

(21) $\cot^{-1}(-1)$

(22) $\csc^{-1}(\text{undefined})$

(23) $\sec^{-1}(1)$

(24) $\cot^{-1}(\text{undefined})$

Instructions: Determine the answers to each problem. Most of the problems have two answers (in two different Quadrants). Check your answers in the back of the book.

(1) $\sin^{-1}\left(-\frac{\sqrt{2}}{2}\right)$ (2) $\cot^{-1}(-1)$ (3) $\csc^{-1}(1)$

(4) $\sin^{-1}(1)$ (5) $\sin^{-1}(0)$ (6) $\tan^{-1}(-1)$

(7) $\csc^{-1}(-1)$ (8) $\sin^{-1}\left(\frac{1}{2}\right)$ (9) $\cot^{-1}(-\sqrt{3})$

(10) $\csc^{-1}(-2)$ (11) $\sec^{-1}(-2)$ (12) $\cot^{-1}(1)$

(13) $\cot^{-1}(-\sqrt{3})$ (14) $\sin^{-1}\left(-\frac{\sqrt{2}}{2}\right)$ (15) $\sin^{-1}\left(-\frac{\sqrt{3}}{2}\right)$

(16) $\tan^{-1}\left(\frac{\sqrt{3}}{3}\right)$ (17) $\csc^{-1}(\sqrt{2})$ (18) $\cot^{-1}(\text{undefined})$

(19) $\tan^{-1}\left(-\frac{\sqrt{3}}{3}\right)$ (20) $\cot^{-1}(0)$ (21) $\sin^{-1}(0)$

(22) $\sin^{-1}\left(-\frac{\sqrt{2}}{2}\right)$ (23) $\sin^{-1}\left(\frac{1}{2}\right)$ (24) $\csc^{-1}\left(-\frac{2\sqrt{3}}{3}\right)$

Instructions: Determine the answers to each problem. Most of the problems have two answers (in two different Quadrants). Check your answers in the back of the book.

(1) $\sec^{-1}(\text{undefined})$

(2) $\tan^{-1}\left(\frac{\sqrt{3}}{3}\right)$

(3) $\sec^{-1}\left(\frac{2\sqrt{3}}{3}\right)$

(4) $\cos^{-1}\left(-\frac{1}{2}\right)$

(5) $\sec^{-1}(2)$

(6) $\cos^{-1}(1)$

(7) $\tan^{-1}(1)$

(8) $\sec^{-1}(-2)$

(9) $\cos^{-1}\left(\frac{\sqrt{2}}{2}\right)$

(10) $\csc^{-1}(-2)$

(11) $\tan^{-1}(-1)$

(12) $\sec^{-1}\left(-\frac{2\sqrt{3}}{3}\right)$

(13) $\tan^{-1}(0)$

(14) $\cos^{-1}(1)$

(15) $\sin^{-1}(0)$

(16) $\sec^{-1}(-2)$

(17) $\cot^{-1}(-1)$

(18) $\tan^{-1}(1)$

(19) $\cos^{-1}\left(-\frac{1}{2}\right)$

(20) $\sin^{-1}(0)$

(21) $\sin^{-1}\left(\frac{\sqrt{3}}{2}\right)$

(22) $\cot^{-1}(0)$

(23) $\tan^{-1}\left(-\frac{\sqrt{3}}{3}\right)$

(24) $\sec^{-1}(-\sqrt{2})$

Instructions: Determine the answers to each problem. Most of the problems have two answers (in two different Quadrants). Check your answers in the back of the book.

(1) $\sin^{-1}\left(\frac{\sqrt{2}}{2}\right)$

(2) $\cot^{-1}(\sqrt{3})$

(3) $\sin^{-1}\left(\frac{1}{2}\right)$

(4) $\tan^{-1}(0)$

(5) $\sin^{-1}\left(\frac{1}{2}\right)$

(6) $\sin^{-1}\left(-\frac{1}{2}\right)$

(7) $\sec^{-1}(2)$

(8) $\sec^{-1}(-1)$

(9) $\cot^{-1}(\text{undefined})$

(10) $\sec^{-1}(2)$

(11) $\sin^{-1}\left(\frac{1}{2}\right)$

(12) $\cos^{-1}(0)$

(13) $\cos^{-1}\left(\frac{\sqrt{3}}{2}\right)$

(14) $\sec^{-1}(2)$

(15) $\tan^{-1}(0)$

(16) $\cot^{-1}(\text{undefined})$

(17) $\sin^{-1}\left(-\frac{\sqrt{3}}{2}\right)$

(18) $\sec^{-1}\left(\frac{2\sqrt{3}}{3}\right)$

(19) $\cos^{-1}\left(\frac{\sqrt{2}}{2}\right)$

(20) $\sec^{-1}(\sqrt{2})$

(21) $\cos^{-1}\left(-\frac{1}{2}\right)$

(22) $\sin^{-1}\left(-\frac{\sqrt{2}}{2}\right)$

(23) $\cos^{-1}(-1)$

(24) $\sin^{-1}\left(\frac{1}{2}\right)$

Chapter 9: The Law of Sines and the Law of Cosines

The law of sines and law of cosines are especially useful for solving for unknown angles and sides in acute (all angles less than 90°) and obtuse (one angle greater than 90°) triangles – i.e. triangles that do not have right (90°) angles. First, we will derive these laws, and then we will discuss how to apply them.

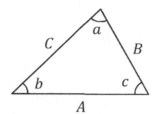   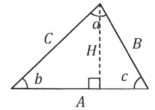

Consider the triangle illustrated above on the left (which is acute, but the resulting laws that we derive will apply in general). In our notation, the uppercase letters A, B, and C represent the lengths of the sides, while the corresponding lowercase letters a, b, and c represent the angles opposite to the respective sides.

The same triangle is also illustrated above to the right, with the addition of a dashed vertical line to represent the triangle's height, H. The height divides the original triangle into two right triangles. Let us write down an equation for the sines of angles b and c:

$$\sin b = \frac{H}{C} \quad , \quad \sin c = \frac{H}{B}$$

We can eliminate the height, H, by solving for it algebraically in each equation:

$$H = C \sin b = B \sin c$$

We can rewrite this as

$$\frac{\sin b}{B} = \frac{\sin c}{C}$$

We can easily rotate the triangle so that a different side is the base and show that, in general,

$$\frac{\sin a}{A} = \frac{\sin b}{B} = \frac{\sin c}{C}$$

The previous equation is known as the law of sines, and is useful for solving for an unknown side or angle in an acute or obtuse triangle, as we will see in the examples.

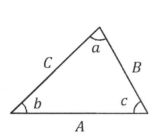 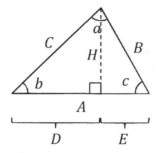

Now let us consider a different property of the same triangle. In the triangle illustrated above on the right, we have divided the base, A, into segments D and E. First, we apply the Pythagorean theorem to each of the right triangles:

$$D^2 + H^2 = C^2 \quad , \quad E^2 + H^2 = B^2$$

Again, we eliminate the height, H, by solving for it in each equation:

$$H^2 = C^2 - D^2 = B^2 - E^2$$

Next, we write down an equation for the cosine of angle c:

$$\cos c = \frac{E}{B}$$

Solving for E in the above equations gives

$$E = B \cos c$$

Looking at the above triangle, it is easy to see that

$$A = D + E$$

Let us substitute $E = B \cos c$ and $D = A - E = A - B \cos c$ into the previous equation with the squares:

$$C^2 - (A - B \cos c)^2 = B^2 - (B \cos c)^2$$

Now we simplify the previous equation algebraically:

124

$$C^2 = (A - B \cos c)^2 + B^2 - (B \cos c)^2$$
$$C^2 = A^2 - 2AB \cos c + B^2 \cos^2 c + B^2 - B^2 \cos^2 c$$

Note that $\cos^2 c$ means to first find the cosine of the angle c, and then square the result. The $+B^2 \cos^2 c$ and $-B^2 \cos^2 c$ terms cancel, yielding the following result:

$$C^2 = A^2 - 2AB \cos c + B^2$$

This equation is known as the law of cosines, and is useful for solving for an unknown side or angle in an acute or obtuse triangle, as we will see in the examples.

Conceptually, the law of cosines is a generalization of the Pythagorean theorem to triangles in general (not just right triangles): The $-2AB \cos c$ term accounts for how the shape of the triangle affects the length C. When $c = 90°$, the triangle is a right triangle, the cross term is zero since $\cos c = \cos 90° = 0$, and the law of cosines becomes the Pythagorean theorem, $C^2 = A^2 + B^2$.

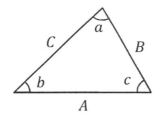

It is also useful to remember that the three internal angles of any triangle always add up to 180°. That is, $a + b + c = 180°$. Once you know two of the three angles, you can solve for the third angle using this relationship.

Take some time to understand these concepts, and study the examples. Once you understand the following examples, you are ready to practice the technique yourself. You may need to refer to the examples frequently as you begin, but should try to solve the exercises all by yourself once you get the hang of it. Be sure to check the answers at the back of the book to ensure that you are solving the problems correctly.

Equations: The law of sines is

$$\frac{\sin a}{A} = \frac{\sin b}{B} = \frac{\sin c}{C}$$

The law of cosines can be written three different ways: $C^2 = A^2 - 2AB \cos c + B^2$, $B^2 = A^2 - 2AC \cos b + C^2$, and $A^2 = B^2 - 2BC \cos a + C^2$. The three interior angles satisfy $a + b + c = 180°$.

Instructions: For each triangle, determine the indicated unknown distance or angle. An unknown angle will be called θ, while an unknown distance will be called x.

Procedure: Apply the law of sines or cosines to solve for the indicated unknown. If you know two of the interior angles, use $a + b + c = 180°$ to solve for the third interior angle.

Label the angles as a, b, and c and their respective opposite sides as A, B, and C. That is, side A must be opposite to angle a, for example. It is very important that you correctly identify the symbols when applying the law of sines and law of cosines.

If you know two sides and the angle opposite to one of the sides, the law of sines can be used to solve for the angle opposite to the other of the two sides that you know. For example, if you know A, B, and a, the law of sines allows you to solve for angle b. Similarly, if you know A, C, and a, the law of sines allows you to solve for c. You can also use the law of sines if you know two angles and one of their opposite sides. For example, if you know A, a, and b, the law of sines allows you to solve for the side B.

If you know two sides and one angle, the law of cosines allows you to solve for the missing side. For example, if you know A, B, and c, the law of cosines allows you to solve for C. If you know all three sides, the law of cosines lets you solve for the interior angles. For example, if you know A, B, and C, the law of cosines allows you to solve for c.

Note that you can use the law of cosines three different ways:

$$C^2 = A^2 - 2AB\cos c + B^2 \quad , \quad B^2 = A^2 - 2AC\cos b + C^2 \quad , \quad A^2 = B^2 - 2BC\cos a + C^2$$

Factor any perfect squares out of the squareroot. For example, in $\sqrt{12}$, we can factor 12 as 3 times 4. Since 4 is a perfect square ($2 \times 2 = 4$), we can write $\sqrt{12} = \sqrt{3 \times 4} = \sqrt{3}\sqrt{4} = 2\sqrt{3}$.

Example 1: $x = ?$ First, label A, B, C, a, b, and c, as we did on the right diagram.

We know a, b, and A, and we're looking for B, so we use the law of sines:
$$\frac{\sin a}{A} = \frac{\sin b}{B} \quad \text{or} \quad \frac{\sin 60°}{9} = \frac{\sin 45°}{x}$$
Cross multiply and then use algebra to solve for the unknown, x:
$$x\sin 60° = 9\sin 45°$$
$$x = 9\frac{\sin 45°}{\sin 60°} = 9\frac{\sqrt{2}/2}{\sqrt{3}/2} = 9\frac{\sqrt{2}}{2}\frac{2}{\sqrt{3}} = \frac{9\sqrt{2}}{\sqrt{3}} = \frac{9\sqrt{2}}{\sqrt{3}}\frac{\sqrt{3}}{\sqrt{3}} = \frac{9\sqrt{6}}{3} = 3\sqrt{6}$$
To divide two fractions, multiply the first fraction by the reciprocal of the second fraction.

126

Example 2: $x = ?$ First, label $A, B, C, a, b,$ and $c,$ as we did on the right diagram.

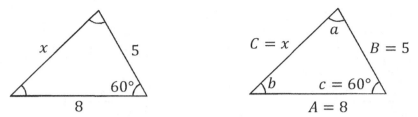

We know $A, B,$ and $c,$ and we're looking for $C,$ so we use the law of cosines:
$$C^2 = A^2 - 2AB \cos c + B^2 \quad \text{or} \quad x^2 = 8^2 - 2(8)(5) \cos 60° + 5^2$$
$$x^2 = 64 - 80\frac{1}{2} + 25 = 64 - 40 + 25 = 49$$
$$x = 7$$

Example 3: $\theta = ?$ First, label $A, B, C, a, b,$ and $c,$ as we did on the right diagram.

We know $B, C,$ and $c,$ and we're looking for $b,$ so we use the law of sines:
$$\frac{\sin b}{B} = \frac{\sin c}{C} \quad \text{or} \quad \frac{\sin \theta}{15} = \frac{\sin 120°}{15\sqrt{3}}$$

Cross multiply and then isolate $\sin \theta$:
$$15\sqrt{3} \sin \theta = 15 \sin 120°$$
$$\sin \theta = \frac{15}{15\sqrt{3}} \frac{\sqrt{3}}{2} = \frac{1}{2}$$

Now take an inverse sine on both sides of the equation:
$$\sin^{-1}(\sin \theta) = \sin^{-1}\left(\frac{1}{2}\right)$$
$$\theta = 30°$$

(When taking an inverse sine, there are two possible answers: The alternate answer, 150°, is not possible since $a + b + c = 180°$ and $c = 120°$.)

Example 4: $\theta = ?$ First, label A, B, C, a, b, and c, as we did on the right diagram.

We know A, B, and C, and we're looking for a, so we use the law of cosines:

$$A^2 = B^2 - 2BC \cos a + C^2 \quad \text{or} \quad (\sqrt{10})^2 = (\sqrt{2})^2 - 2(\sqrt{2})(4)\cos\theta + 4^2$$

Use algebra to isolate $\cos\theta$:

$$10 = 2 - 8\sqrt{2}\cos\theta + 16$$

$$8\sqrt{2}\cos\theta = 2 + 16 - 10 = 8$$

$$\cos\theta = \frac{8}{8\sqrt{2}} = \frac{1}{\sqrt{2}}\frac{\sqrt{2}}{\sqrt{2}} = \frac{\sqrt{2}}{2}$$

Now take an inverse cosine on both sides of the equation:

$$\theta = \cos^{-1}(\cos\theta) = \cos^{-1}\left(\frac{\sqrt{2}}{2}\right) = 45°$$

(The alternate answer, 315°, to the inverse cosine is not possible since $a + b + c = 180°$.)

Example 5: $x = ?$ First, label A, B, C, a, b, and c, as we did on the right diagram.

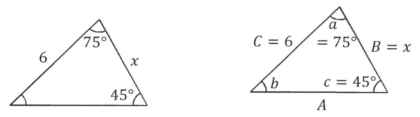

We know a, c, and C, and we're looking for B. We could use the law of sines if we knew b. So we first find b, knowing that the interior angles add up to 180°: $b = 180° - a - c = 180° - 75° - 45° = 60°$. Now we use the law of sines:

$$\frac{\sin b}{B} = \frac{\sin c}{C} \quad \text{or} \quad \frac{\sin 60°}{x} = \frac{\sin 45°}{6}$$

Cross multiply and then use algebra to solve for the unknown, x:

$$x \sin 45° = 6 \sin 60°$$

$$x = 6\frac{\sin 60°}{\sin 45°} = 6\frac{\sqrt{3}/2}{\sqrt{2}/2} = 6\frac{\sqrt{3}}{2}\frac{2}{\sqrt{2}} = \frac{6\sqrt{3}}{\sqrt{2}} = \frac{6\sqrt{3}}{\sqrt{2}}\frac{\sqrt{2}}{\sqrt{2}} = \frac{6\sqrt{6}}{2} = 3\sqrt{6}$$

To divide two fractions, multiply the first fraction by the reciprocal of the second fraction.

128

Example 6: $x = ?$ First, label $A, B, C, a, b,$ and $c,$ as we did on the right diagram.

We know $A, C,$ and $c,$ and we're looking for $B,$ so we use the law of cosines:

$$C^2 = A^2 - 2AB \cos c + B^2 \quad \text{or} \quad 5^2 = \left(\sqrt{5}\right)^2 - 2\left(\sqrt{5}\right)x \cos 135° + x^2$$

$$25 = 5 - 2\sqrt{5}x\left(-\frac{\sqrt{2}}{2}\right) + x^2$$

$$20 = x\sqrt{10} + x^2$$

Since there is a term proportional to $x^2,$ a term proportional to $x,$ and a constant term, this is the quadratic equation. The first step toward solving the quadratic equation is to cast the equation in standard form, which is $ax^2 + bx + c = 0$ (where the constants $a, b,$ and c are the standard symbols for the quadratic equation – they are not the angles from the triangle above, even though the same symbols are used for both). To do this, we move the terms around (following the rules of algebra, of course):

$$x^2 + x\sqrt{10} - 20 = 0$$

We can now identify the constants of the quadratic equation:

$$a = 1 \quad , \quad b = \sqrt{10} \quad , \quad c = -20$$

The solution to the quadratic equation, $ax^2 + bx + c = 0,$ is:

$$x = \frac{-b \pm \sqrt{b^2 - 4ac}}{2a}$$

Substituting the constants $a, b,$ and c into the quadratic formula gives

$$x = \frac{-\sqrt{10} \pm \sqrt{\left(\sqrt{10}\right)^2 - 4(1)(-20)}}{2(1)}$$

$$x = \frac{-\sqrt{10} \pm \sqrt{10 + 80}}{2} = \frac{-\sqrt{10} \pm \sqrt{90}}{2}$$

Here, we factor out the $\sqrt{10}$:

$$x = \sqrt{10}\left(\frac{-1 \pm \sqrt{9}}{2}\right) = \sqrt{10}\left(\frac{-1 \pm 3}{2}\right)$$

Since x is the length of a triangle, it can't be negative, so we only keep the root that makes x positive:

$$x = \sqrt{10}\left(\frac{-1 + 3}{2}\right) = \sqrt{10}\left(\frac{2}{2}\right) = \sqrt{10}$$

Instructions: For each right triangle, determine the indicated unknown distance or angle. Check your answers in the back of the book. No calculator is needed.

Note: The triangles are <u>not drawn to scale</u>. Except for obtuse triangles where the largest angle and longest side are, in fact, largest, a side that looks longer than another may actually be shorter and an angle that looks larger than another may actually be smaller.

(1) $x = ?$

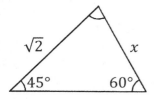

(2) $x = ?$

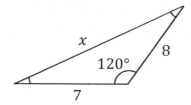

(3) $\theta = ?$

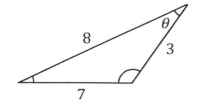

(4) $\theta = ?$

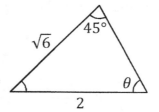

Instructions: For each right triangle, determine the indicated unknown distance or angle. Check your answers in the back of the book. No calculator is needed.

Note: The triangles are <u>not drawn to scale</u>. Except for obtuse triangles where the largest angle and longest side are, in fact, largest, a side that looks longer than another may actually be shorter and an angle that looks larger than another may actually be smaller.

(1) $x = ?$

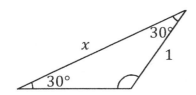

(2) $x = ?$

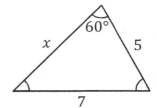

(3) $\theta = ?$

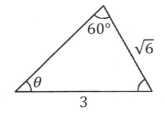

(4) $x = ?$

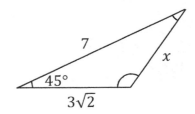

Instructions: For each right triangle, determine the indicated unknown distance or angle. Check your answers in the back of the book. No calculator is needed.

Note: The triangles are <u>not drawn to scale</u>. Except for obtuse triangles where the largest angle and longest side are, in fact, largest, a side that looks longer than another may actually be shorter and an angle that looks larger than another may actually be smaller.

(1) $\theta = ?$

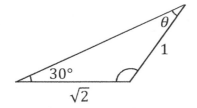

(2) $\theta = ?$

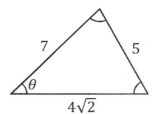

(3) $x = ?$

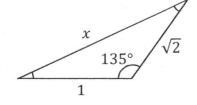

(4) $x = ?$

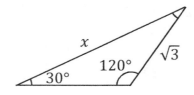

Instructions: For each right triangle, determine the indicated unknown distance or angle. Check your answers in the back of the book. No calculator is needed.

Note: The triangles are <u>not drawn to scale</u>. Except for obtuse triangles where the largest angle and longest side are, in fact, largest, a side that looks longer than another may actually be shorter and an angle that looks larger than another may actually be smaller.

(1) $x = ?$

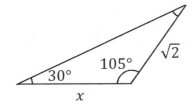

(2) $x = ?$

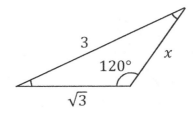

(3) $\theta = ?$

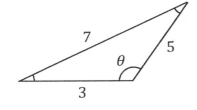

(4) $x = ?$

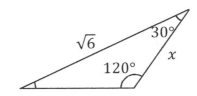

Instructions: For each right triangle, determine the indicated unknown distance or angle. Check your answers in the back of the book. No calculator is needed.

Note: The triangles are not drawn to scale. Except for obtuse triangles where the largest angle and longest side are, in fact, largest, a side that looks longer than another may actually be shorter and an angle that looks larger than another may actually be smaller.

(1) $x = ?$

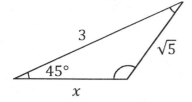

(2) $\theta = ?$

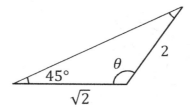

(3) $\theta = ?$

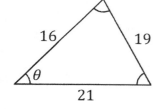

(4) $x = ?$

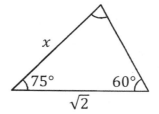

Instructions: For each right triangle, determine the indicated unknown distance or angle. Check your answers in the back of the book. No calculator is needed.

Note: The triangles are <u>not drawn to scale</u>. Except for obtuse triangles where the largest angle and longest side are, in fact, largest, a side that looks longer than another may actually be shorter and an angle that looks larger than another may actually be smaller.

(1) $x = ?$

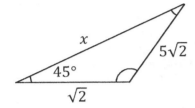

(2) $x = ?$

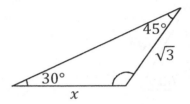

(3) $\theta = ?$

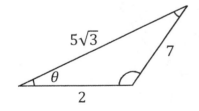

(4) $x = ?$

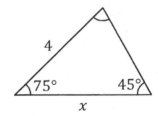

Instructions: For each right triangle, determine the indicated unknown distance or angle. Check your answers in the back of the book. No calculator is needed.

Note: The triangles are <u>not drawn to scale</u>. Except for obtuse triangles where the largest angle and longest side are, in fact, largest, a side that looks longer than another may actually be shorter and an angle that looks larger than another may actually be smaller.

(1) $\theta = ?$

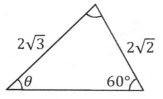

(2) $x = ?$

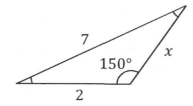

(3) $\theta = ?$

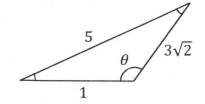

(4) $\theta = ?$

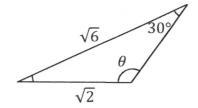

Instructions: For each right triangle, determine the indicated unknown distance or angle. Check your answers in the back of the book. No calculator is needed.

Note: The triangles are <u>not drawn to scale</u>. Except for obtuse triangles where the largest angle and longest side are, in fact, largest, a side that looks longer than another may actually be shorter and an angle that looks larger than another may actually be smaller.

(1) $x = ?$

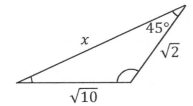

(2) $x = ?$

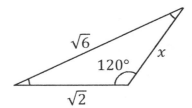

(3) $\theta = ?$

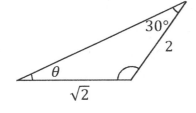

(4) $x = ?$

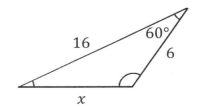

Instructions: For each right triangle, determine the indicated unknown distance or angle. Check your answers in the back of the book. No calculator is needed.

Note: The triangles are <u>not drawn to scale</u>. Except for obtuse triangles where the largest angle and longest side are, in fact, largest, a side that looks longer than another may actually be shorter and an angle that looks larger than another may actually be smaller.

(1) $x = ?$

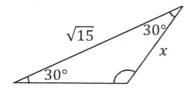

(2) $x = ?$

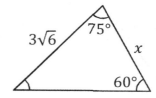

(3) $\theta = ?$

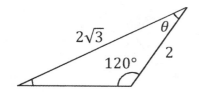

(4) $x = ?$

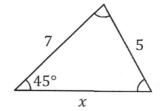

Instructions: For each right triangle, determine the indicated unknown distance or angle. Check your answers in the back of the book. No calculator is needed.

Note: The triangles are <u>not drawn to scale</u>. Except for obtuse triangles where the largest angle and longest side are, in fact, largest, a side that looks longer than another may actually be shorter and an angle that looks larger than another may actually be smaller.

(1) $x = ?$

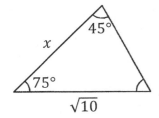

(2) $x = ?$

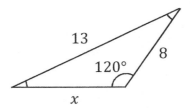

(3) $x = ?$

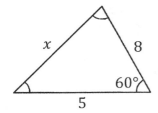

(4) $x = ?$

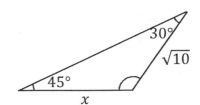

Instructions: For each right triangle, determine the indicated unknown distance or angle. Check your answers in the back of the book. No calculator is needed.

Note: The triangles are <u>not drawn to scale</u>. Except for obtuse triangles where the largest angle and longest side are, in fact, largest, a side that looks longer than another may actually be shorter and an angle that looks larger than another may actually be smaller.

(1) $x = ?$

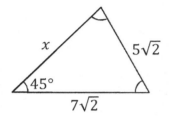

(2) $x = ?$

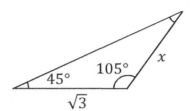

(3) $\theta = ?$

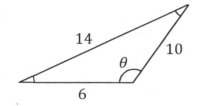

(4) $\theta = ?$

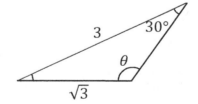

Instructions: For each right triangle, determine the indicated unknown distance or angle. Check your answers in the back of the book. No calculator is needed.

Note: The triangles are <u>not drawn to scale</u>. Except for obtuse triangles where the largest angle and longest side are, in fact, largest, a side that looks longer than another may actually be shorter and an angle that looks larger than another may actually be smaller.

(1) $x = ?$

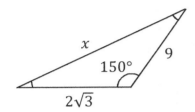

(2) $\theta = ?$

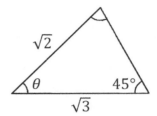

(3) $x = ?$

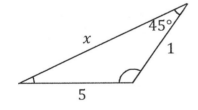

(4) $\theta = ?$

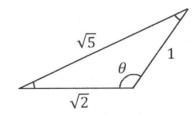

Chapter 10: Learn and Apply Trig Identities

There are several common and useful trigonometric identities, which can be helpful for solving a variety of problems and for deriving useful formulas in math and science. We will first derive several trigonometric identities, and then focus on how to apply them to solve problems or derive other formulas.

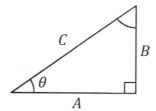

The Pythagorean identities are trigonometric versions of the Pythagorean theorem. For the right triangle illustrated above, the Pythagorean theorem is

$$A^2 + B^2 = C^2$$

Since $\sin \theta = B/C$, we can write $B = C \sin \theta$. Similarly, since $\cos \theta = A/C$, we can write $A = C \cos \theta$. We substitute these expressions into the Pythagorean theorem:

$$(C \cos \theta)^2 + (C \sin \theta)^2 = C^2$$
$$C^2 \cos^2 \theta + C^2 \sin^2 \theta = C^2$$

Note, for example, that $\cos^2 \theta$ means to first find the cosine of the angle θ, and then square the result. Cancelling out the C^2, which is common to each term, we find that

$$\cos^2 \theta + \sin^2 \theta = 1$$

We can obtain a second Pythagorean identity by dividing this equation by $\cos^2 \theta$:

$$\frac{\cos^2 \theta}{\cos^2 \theta} + \frac{\sin^2 \theta}{\cos^2 \theta} = \frac{1}{\cos^2 \theta}$$
$$1 + \tan^2 \theta = \sec^2 \theta$$

A third Pythagorean identity results by dividing the original Pythagorean identity by $\sin^2 \theta$:

$$\frac{\cos^2\theta}{\sin^2\theta} + \frac{\sin^2\theta}{\sin^2\theta} = \frac{1}{\sin^2\theta}$$
$$\cot^2\theta + 1 = \csc^2\theta$$

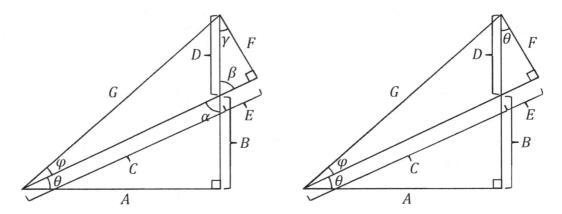

We will use the diagram above to prove the angle sum identities. Note that the same diagram appears on the left and right; the difference is that gamma (γ) is labeled as theta (θ), and alpha (α) and beta (β) have been removed. Note that $\alpha = 90° - \theta$ since $\alpha + \theta + 90° = 180°$. Also, $\beta = \alpha$ because they are vertical angles. Furthermore, $\gamma = 90° - \beta$ since $\beta + \gamma + 90° = 180°$. This mean that $\gamma = 90° - \alpha$ (since $\beta = \alpha$). Compare this with $\theta = 90° - \alpha$ (which comes from the original equation, $\alpha = 90° - \theta$). Thus, we have proven that $\theta = \gamma$. This is why the diagram to the right has θ in place of γ. Henceforth, we will use the diagram on the right to prove the angle sum identities.

First, we express sine and cosine for each labeled angle in the four right triangles:

$$\sin\varphi = \frac{F}{G}, \quad \sin\theta = \frac{B}{C} = \frac{E}{D}, \quad \sin(\theta + \varphi) = \frac{B + D}{G}$$
$$\cos\varphi = \frac{C + E}{G}, \quad \cos\theta = \frac{A}{C} = \frac{F}{D}, \quad \cos(\theta + \varphi) = \frac{A}{G}$$

Think through the following combination of trig functions:

$$\sin\theta\cos\varphi + \cos\theta\sin\varphi = \left(\frac{E}{D}\right)\left(\frac{C + E}{G}\right) + \left(\frac{F}{D}\right)\left(\frac{F}{G}\right) = \frac{EC + E^2 + F^2}{DG}$$

Note that $E^2 + F^2 = D^2$ according to the Pythagorean theorem. Also, since $\frac{B}{C} = \frac{E}{D}$ (see the $\sin\theta$ formulas above), if we cross multiply we get $EC = BD$. Now plug $E^2 + F^2 = D^2$ and $EC = BD$ into the previous equation:

$$\sin\theta\cos\varphi + \cos\theta\sin\varphi = \frac{BD + D^2}{DG} = \frac{(B + D)D}{DG} = \frac{B + D}{G} = \sin(\theta + \varphi)$$

143

We have just derived the angle sum identity for sine:

$$\sin(\theta + \varphi) = \sin\theta\cos\varphi + \cos\theta\sin\varphi$$

Note that $\sin(\theta + \varphi)$ means to find the sine of $\theta + \varphi$, and that sine is a function; $\sin(\theta + \varphi)$ is not a multiplication, and so it is not distributive (i.e. it does **not** equal $\sin\theta + \sin\varphi$).
We now proceed to derive another angle sum identity:

$$\cos\theta\cos\varphi - \sin\theta\sin\varphi = \left(\frac{F}{D}\right)\left(\frac{C+E}{G}\right) - \left(\frac{E}{D}\right)\left(\frac{F}{G}\right) = \frac{FC + EF - EF}{DG} = \frac{FC}{DG}$$

Since $\frac{A}{C} = \frac{F}{D}$ (see the $\cos\theta$ formulas on the previous page), it follows that $\frac{C}{D} = \frac{A}{F}$ by algebraic rearrangement. Substituting this into the previous equation, we get

$$\cos\theta\cos\varphi - \sin\theta\sin\varphi = \frac{F\,C}{G\,D} = \frac{F\,A}{G\,F} = \frac{A}{G} = \cos(\theta + \varphi)$$

This is the angle sum identity for cosine:

$$\cos(\theta + \varphi) = \cos\theta\cos\varphi - \sin\theta\sin\varphi$$

In order to derive the angle sum identity for tangent, we divide sine by cosine:

$$\tan(\theta + \varphi) = \frac{\sin(\theta + \varphi)}{\cos(\theta + \varphi)} = \frac{\sin\theta\cos\varphi + \cos\theta\sin\varphi}{\cos\theta\cos\varphi - \sin\theta\sin\varphi}$$

Next, we divide both the numerator and denominator of this fraction by $\cos\theta\cos\varphi$:

$$\tan(\theta + \varphi) = \frac{\dfrac{\sin\theta\cos\varphi + \cos\theta\sin\varphi}{\cos\theta\cos\varphi}}{\dfrac{\cos\theta\cos\varphi - \sin\theta\sin\varphi}{\cos\theta\cos\varphi}} = \frac{\tan\theta + \tan\varphi}{1 - \tan\theta\tan\varphi}$$

The angle difference identities can be derived from the angle sum identities by noting that sine is an odd function, while cosine is an even function. That is,

$$\sin(-\theta) = -\sin\theta \quad , \quad \cos(-\theta) = \cos\theta$$

This should make sense based on how we defined the reference angle and how we used the reference angle to find the sine and cosine in other Quadrants. Sine is positive in Quadrants I and II, and negative in Quadrants III and IV; cosine is positive in Quadrants I and IV, and negative in Quadrants II and III.

144

Note that if θ lies in Quadrant I, then $-\theta$ lies in Quadrant IV; and if θ lies in Quadrant II, then $-\theta$ lies in Quadrant III; and so on. Cosine is positive in both Quadrants I and IV, so swapping sign from θ to $-\theta$ leaves cosine unchanged; similarly, cosine is negative in both Quadrants II and III. Sine is positive in Quadrant I, but negative in Quadrant IV, so swapping sign from θ to $-\theta$ changes the sign of sine; similarly, sine has opposite sign in Quadrants II and III. Note that the reference angle is the same for θ as it is for $-\theta$. Since tangent equals sine over cosine, and since sine is odd while cosine is even, tangent is an odd function:

$$\tan(-\theta) = -\tan\theta$$

We simply change the sign of φ in the angle sum identities – and use $\sin(-\theta) = -\sin\theta$ and $\cos(-\theta) = \cos\theta$ – to derive the angle difference identities:

$$\sin(\theta - \varphi) = \sin\theta\cos(-\varphi) + \cos\theta\sin(-\varphi) = \sin\theta\cos\varphi - \cos\theta\sin\varphi$$
$$\cos(\theta - \varphi) = \cos\theta\cos(-\varphi) - \sin\theta\sin(-\varphi) = \cos\theta\cos\varphi + \sin\theta\sin\varphi$$
$$\tan(\theta - \varphi) = \frac{\tan\theta + \tan(-\varphi)}{1 - \tan\theta\tan(-\varphi)} = \frac{\tan\theta - \tan\varphi}{1 + \tan\theta\tan\varphi}$$

The angle sum and difference identities can be combined together as follows:

$$\sin(\theta \pm \varphi) = \sin\theta\cos\varphi \pm \cos\theta\sin\varphi$$
$$\cos(\theta \pm \varphi) = \cos\theta\cos\varphi \mp \sin\theta\sin\varphi$$
$$\tan(\theta \pm \varphi) = \frac{\tan\theta \pm \tan\varphi}{1 \mp \tan\theta\tan\varphi}$$

Note that upper signs and lower signs all go together (i.e. when one sign is upper, all are upper).

Next, we will derive the double-angle identities. We do this by setting $\varphi = \theta$ in the angle sum identities:

$$\sin(2\theta) = \sin(\theta + \theta) = \sin\theta\cos\theta + \cos\theta\sin\theta = 2\sin\theta\cos\theta$$
$$\cos(2\theta) = \cos(\theta + \theta) = \cos\theta\cos\theta - \sin\theta\sin\theta = \cos^2\theta - \sin^2\theta$$
$$\tan(2\theta) = \tan(\theta + \theta) = \frac{\tan\theta + \tan\theta}{1 - \tan\theta\tan\theta} = \frac{2\tan\theta}{1 - \tan^2\theta}$$

In addition to double-angle identities, there are also half-angle identities. In order to derive the half-angle identities, we begin by rewriting the double-angle identity for cosine as follows:

$$\cos\theta = \cos^2\left(\frac{\theta}{2}\right) - \sin^2\left(\frac{\theta}{2}\right)$$

Note that we have simply changed our perspective, since we aim to derive a half-angle formula instead of a double-angle formula. We use the Pythagorean identity (namely, since $\sin^2\theta + \cos^2\theta = 1$, $\cos^2\theta = 1 - \sin^2\theta$) to rewrite the above equation as

$$\cos\theta = 1 - \sin^2\left(\frac{\theta}{2}\right) - \sin^2\left(\frac{\theta}{2}\right) = 1 - 2\sin^2\left(\frac{\theta}{2}\right)$$

Now we solve for the sine function:

$$\sin\left(\frac{\theta}{2}\right) = \pm\sqrt{\frac{1 - \cos\theta}{2}}$$

The $\pm$ in the half-angle formula for sine reflects that the squareroot can be positive or negative. We can derive the half-angle formula for cosine by instead making the substitution $\sin^2\theta = 1 - \cos^2\theta$ where we had made the substitution $\cos^2\theta = 1 - \sin^2\theta$ (here, you must be careful to distribute correctly):

$$\cos\theta = \cos^2\left(\frac{\theta}{2}\right) - 1 + \cos^2\left(\frac{\theta}{2}\right) = 2\cos^2\left(\frac{\theta}{2}\right) - 1$$

$$\cos\left(\frac{\theta}{2}\right) = \pm\sqrt{\frac{1 + \cos\theta}{2}}$$

We obtain the half-angle identity for tangent by dividing sine by cosine:

$$\tan\left(\frac{\theta}{2}\right) = \pm\sqrt{\frac{1 - \cos\theta}{1 + \cos\theta}}$$

There are other trig identities, but these are the most common. These identities are used commonly when trig is involved in science courses like physics and in higher-level math courses. Other trig identities can be derived from these – such as trig identities involving secant, cosecant, and cotangent.

We will now summarize these trig identities and then focus on how to apply the trig identities to solve problems and to derive other formulas.

Take some time to understand these concepts, and study the examples. Once you understand the following examples, you are ready to practice the technique yourself. You may need to refer to the examples frequently as you begin, but should try to solve the exercises all by yourself once you get the hang of it. Be sure to check the answers at the back of the book to ensure that you are solving the problems correctly.

Trig identities: Here is a convenient table of the trig identities that we derived:

$$\cos^2 \theta + \sin^2 \theta = 1$$
$$1 + \tan^2 \theta = \sec^2 \theta \quad \text{(Pythagorean identities)}$$
$$\cot^2 \theta + 1 = \csc^2 \theta$$

$$\sin(\theta \pm \varphi) = \sin \theta \cos \varphi \pm \cos \theta \sin \varphi$$
$$\cos(\theta \pm \varphi) = \cos \theta \cos \varphi \mp \sin \theta \sin \varphi$$
$$\tan(\theta \pm \varphi) = \frac{\tan \theta \pm \tan \varphi}{1 \mp \tan \theta \tan \varphi} \quad \text{(angle sum/difference identities)}$$

$$\sin(-\theta) = -\sin \theta \quad , \quad \cos(-\theta) = \cos \theta \quad , \quad \tan(-\theta) = -\tan \theta \quad \text{(odd/even)}$$

$$\sin(2\theta) = 2 \sin \theta \cos \theta$$
$$\cos(2\theta) = \cos^2 \theta - \sin^2 \theta \quad \text{(double-angle identities)}$$
$$\tan(2\theta) = \frac{2 \tan \theta}{1 - \tan^2 \theta}$$

$$\sin\left(\frac{\theta}{2}\right) = \pm \sqrt{\frac{1 - \cos \theta}{2}}$$

$$\cos\left(\frac{\theta}{2}\right) = \pm \sqrt{\frac{1 + \cos \theta}{2}} \quad \text{(half-angle identities)}$$

$$\tan\left(\frac{\theta}{2}\right) = \pm \sqrt{\frac{1 - \cos \theta}{1 + \cos \theta}}$$

Instructions: There are two types of problems – some are numerical, others are symbolic. For the problems with numbers, select the specified trig identity and plug the numbers in; you must use the specified trig identity, even if the problem can be solved without using it. For the problems with symbols, derive the indicated equation by combining trig identities together through substitution and applying algebra. The problems that ask you to derive an equation do not have answers at the back of the book (since the answer is instead given in the question).

Procedure for numerical problems: Select the specified trig identity and plug numbers into it. In some cases, you may need to first modify the specified trig identity – by combining it together with other equations through substitution. For example, if you need to work with secant, you will first need to modify a trig identity for cosine using the equation $\sec \theta = 1/\cos \theta$. Study the examples before you begin. You must use the specified trig identity, even if the problem can be solved without using it.

Reduce any fractions by dividing both the numerator and denominator by the greatest common factor. Factor any perfect squares out of the squareroot. For example, in $\sqrt{12}$, we can factor 12 as 3 times 4. Since 4 is a perfect square ($2 \times 2 = 4$), we can write $\sqrt{12} = \sqrt{3 \times 4} = \sqrt{3}\sqrt{4} = 2\sqrt{3}$. Also, rationalize the denominator. For example, $\frac{1}{\sqrt{2}} = \frac{1}{\sqrt{2}}\frac{\sqrt{2}}{\sqrt{2}} = \frac{\sqrt{2}}{2}$ and $\frac{6}{\sqrt{3}} = \frac{6}{\sqrt{3}}\frac{\sqrt{3}}{\sqrt{3}} = \frac{6\sqrt{3}}{3} = 2\sqrt{3}$.

Procedure for symbolic problems: Derive the indicated equation by combining trig identities (and, where applicable, the definitions of secant, cosecant, and cotangent) together through substitution and applying algebra. Study the examples before you begin.

Example 1: For $\sin \theta = \frac{1}{4}$, determine $\cos \theta$ using a Pythagorean identity.

Although you can't solve for θ without using a calculator, you can still solve for $\cos \theta$. We begin with the specified identity:
$$\cos^2 \theta + \sin^2 \theta = 1$$
$$\cos \theta = \pm\sqrt{1 - \sin^2 \theta} = \pm\sqrt{1 - \left(\frac{1}{4}\right)^2} = \pm\sqrt{1 - \frac{1}{16}} = \pm\sqrt{\frac{16}{16} - \frac{1}{16}} = \pm\sqrt{\frac{15}{16}} = \pm\frac{\sqrt{15}}{4}$$

Example 2: Use trig identities to show that $\tan^2 \theta + \sec^2 \theta = 2\sec^2 \theta - 1$.

The relevant Pythagorean identity is $1 + \tan^2 \theta = \sec^2 \theta$. Solving for $\tan^2 \theta$ gives $\tan^2 \theta = \sec^2 \theta - 1$. Substitute this into the left-hand side of the indicated equation:
$$\tan^2 \theta + \sec^2 \theta = \sec^2 \theta - 1 + \sec^2 \theta = 2\sec^2 \theta - 1$$

Example 3: Use an angle sum identity to calculate $\sec 75°$.

First find $\cos 75°$, and then reciprocate this to find the secant. We can think of $75°$ as $30° + 45°$:
$$\cos(30° + 45°) = \cos 30° \cos 45° - \sin 30° \sin 45°$$
$$\cos 75° = \left(\frac{\sqrt{3}}{2}\right)\left(\frac{\sqrt{2}}{2}\right) - \left(\frac{1}{2}\right)\left(\frac{\sqrt{2}}{2}\right) = \frac{\sqrt{6}}{4} - \frac{\sqrt{2}}{4} = \frac{\sqrt{6} - \sqrt{2}}{4}$$
Reciprocate the cosine to find the secant:
$$\sec 75° = \frac{1}{\cos 75°} = \frac{4}{\sqrt{6} - \sqrt{2}}\frac{\sqrt{6} + \sqrt{2}}{\sqrt{6} + \sqrt{2}} = \frac{4\sqrt{6} + 4\sqrt{2}}{6 - 2} = \frac{4\sqrt{6} + 4\sqrt{2}}{4} = \sqrt{6} + \sqrt{2}$$

Example 4: Use trig identities to show that $\sin(3\theta) = 3\sin \theta - 4\sin^3 \theta$.

We can think of 3θ as $2\theta + \theta$:
$$\sin(2\theta + \theta) = \sin(2\theta) \cos \theta + \cos(2\theta) \sin \theta$$
$$\sin(3\theta) = (2\sin \theta \cos \theta) \cos \theta + (\cos^2 \theta - \sin^2 \theta) \sin \theta$$
$$\sin(3\theta) = 2\sin \theta \cos^2 \theta + \cos^2 \theta \sin \theta - \sin^3 \theta = 3\sin \theta \cos^2 \theta - \sin^3 \theta$$
$$\sin(3\theta) = 3\sin \theta (1 - \sin^2 \theta) - \sin^3 \theta = 3\sin \theta - 4\sin^3 \theta$$
We used the Pythagorean identity, in the form $\cos^2 \theta = 1 - \sin^2 \theta$, in the last step.

Example 5: Use a double-angle identity to calculate $\tan 120°$, and check your result.

Although we learned how to calculate $\tan 120°$ in Chapter 7, the instructions specifically ask us to use a double-angle identity instead. We can think of $120°$ as $2 \times 60°$:

$$\tan[2(60°)] = \frac{2\tan 60°}{1 - \tan^2(60°)} = \frac{2\sqrt{3}}{1 - \left(\sqrt{3}\right)^2} = \frac{2\sqrt{3}}{1 - 3} = \frac{2\sqrt{3}}{-2} = -\sqrt{3}$$

Using the technique of Chapter 7, you should be able to check that $\tan 120° = -\sqrt{3}$.

Example 6: Use trig identities to show that $\tan\left(\frac{\theta}{2}\right) = \frac{1 - \cos\theta}{\sin\theta}$.

Let us begin with the half-angle identity for tangent:

$$\tan\left(\frac{\theta}{2}\right) = \pm\sqrt{\frac{1 - \cos\theta}{1 + \cos\theta}}$$

This equation has a squareroot, whereas the equation that we are trying to derive does not have a squareroot. The trick is to rationalize the denominator by multiplying the numerator and denominator by the conjugate of the denominator:

$$\tan\left(\frac{\theta}{2}\right) = \pm\sqrt{\left(\frac{1 - \cos\theta}{1 + \cos\theta}\right)\left(\frac{1 - \cos\theta}{1 - \cos\theta}\right)}$$

$$\tan\left(\frac{\theta}{2}\right) = \pm\sqrt{\frac{(1 - \cos\theta)(1 - \cos\theta)}{(1 + \cos\theta)(1 - \cos\theta)}}$$

$$\tan\left(\frac{\theta}{2}\right) = \pm\sqrt{\frac{(1 - \cos\theta)^2}{1 + \cos\theta - \cos\theta - \cos^2\theta}}$$

$$\tan\left(\frac{\theta}{2}\right) = \pm\sqrt{\frac{(1 - \cos\theta)^2}{1 - \cos^2\theta}}$$

$$\tan\left(\frac{\theta}{2}\right) = \pm\sqrt{\frac{(1 - \cos\theta)^2}{\sin^2\theta}}$$

$$\tan\left(\frac{\theta}{2}\right) = \frac{1 - \cos\theta}{\sin\theta}$$

(The signs work themselves out in this case.)

Example 7: Use a half-angle identity to calculate $\sin 22.5°$.

We can think of $22.5°$ as $45°/2$:

$$\sin\left(\frac{45°}{2}\right) = \pm\sqrt{\frac{1 - \cos 45°}{2}} = \pm\sqrt{\frac{1 - \frac{\sqrt{2}}{2}}{2}} = \pm\sqrt{\frac{1}{2} - \frac{\sqrt{2}}{4}} = \pm\sqrt{\frac{2}{4} - \frac{\sqrt{2}}{4}} = \frac{\sqrt{2 - \sqrt{2}}}{2}$$

The positive root is needed since sine is positive for $22.5°$ (Quadrant I).

Instructions: Select the specified trig identity and plug the numbers in. You must use the specified trig identity, even if the problem can be solved without using it. Check your answers in the back of the book. No calculator is needed.

(1) Use an angle sum identity to calculate $\sin 105°$.

(2) Use an angle sum identity to calculate $\tan 165°$.

(3) Use an angle sum identity to calculate $\csc 75°$.

(4) Use an angle sum identity to calculate $\cos 105°$.

(5) Use an angle sum identity to calculate $\cot 195°$.

Instructions: Select the specified trig identity and plug the numbers in. You must use the specified trig identity, even if the problem can be solved without using it. Check your answers in the back of the book. No calculator is needed.

(1) Use an angle sum identity to calculate $\sec 255°$.

(2) Use an angle sum identity to calculate $\sin 285°$.

(3) Use an angle sum identity to calculate $\cos 345°$.

(4) Use an angle sum identity to calculate $\cot 165°$.

(5) Use an angle sum identity to calculate $\csc 285°$.

Instructions: Select the specified trig identity and plug the numbers in. You must use the specified trig identity, even if the problem can be solved without using it. Check your answers in the back of the book. No calculator is needed.

(1) Use an angle difference identity to calculate $\cos 15°$.

(2) Use an angle difference identity to calculate $\tan 75°$.

(3) Use an angle difference identity to calculate $\sin 165°$.

(4) Use an angle difference identity to calculate $\sec 195°$.

(5) Use an angle difference identity to calculate $\cot 345°$.

Instructions: Select the specified trig identity and plug the numbers in. You must use the specified trig identity, even if the problem can be solved without using it. Check your answers in the back of the book. No calculator is needed.

(1) Use an angle difference identity to calculate $\sin 75°$.

(2) Use an angle difference identity to calculate $\sec 255°$.

(3) Use an angle difference identity to calculate $\csc 195°$.

(4) Use an angle difference identity to calculate $\cot 15°$.

(5) Use an angle difference identity to calculate $\tan 285°$.

Instructions: Select the specified trig identity and plug the numbers in. You must use the specified trig identity, even if the problem can be solved without using it. Check your answers in the back of the book. No calculator is needed.

(1) Use a double angle identity to calculate $\tan 240°$.

(2) Use a double angle identity to calculate $\cos 180°$.

(3) Use a double angle identity to calculate $\sec 300°$.

(4) Use a double angle identity to calculate $\cot 270°$.

(5) Use a double angle identity to calculate $\csc 120°$.

Instructions: Select the specified trig identity and plug the numbers in. You must use the specified trig identity, even if the problem can be solved without using it. Check your answers in the back of the book. No calculator is needed.

(1) Use a double-angle identity to calculate $\sin 360°$.

(2) Use a double-angle identity to calculate $\sec 120°$.

(3) Use a double-angle identity to calculate $\cot 180°$.

(4) Use a double-angle identity to calculate $\tan 450°$.

(5) Use a double-angle identity to calculate $\cos 630°$.

Instructions: Select the specified trig identity and plug the numbers in. You must use the specified trig identity, even if the problem can be solved without using it. Check your answers in the back of the book. No calculator is needed.

(1) Use a half-angle identity to calculate cos 22.5°.

(2) Use a half-angle identity to calculate sin 15°.

(3) Use a half-angle identity to calculate tan 67.5°.

(4) Use a half-angle identity to calculate cot 75°.

(5) Use a half-angle identity to calculate sec 112.5°.

Instructions: Select the specified trig identity and plug the numbers in. You must use the specified trig identity, even if the problem can be solved without using it. Check your answers in the back of the book. No calculator is needed.

(1) Use a half-angle identity to calculate $\sin 7.5°$.

(2) Use a half-angle identity to calculate $\cos 37.5°$.

(3) Use a half-angle identity to calculate $\tan 67.5°$.

(4) Use a half-angle identity to calculate $\csc 157.5°$.

(5) Use a half-angle identity to calculate $\cot 75°$.

Instructions: Select the specified trig identity and plug the numbers in. You must use the specified trig identity, even if the problem can be solved without using it. Check your answers in the back of the book. No calculator is needed.

(1) For $\cos\theta = \frac{3}{5}$, determine $\sin\theta$ using a Pythagorean identity.

(2) For $\tan\theta = 2$, determine $\sec\theta$ using a Pythagorean identity.

(3) For $\sin\theta = \frac{12}{13}$, determine $\cos\theta$ using a Pythagorean identity.

(4) For $\csc\theta = 3$, determine $\cot\theta$ using a Pythagorean identity.

(5) For $\cos\theta = \frac{1}{3}$, determine $\sin\theta$ using a Pythagorean identity.

Instructions: Select the specified trig identity and plug the numbers in. You must use the specified trig identity, even if the problem can be solved without using it. Check your answers in the back of the book. No calculator is needed.

(1) For $\sec \theta = 2$, determine $\tan \theta$ using a Pythagorean identity.

(2) For $\sin \theta = \frac{3}{4}$, determine $\cos \theta$ using a Pythagorean identity.

(3) For $\cot \theta = \frac{1}{2}$, determine $\csc \theta$ using a Pythagorean identity.

(4) For $\cos \theta = \frac{2}{3}$, determine $\sin \theta$ using a Pythagorean identity.

(5) For $\tan \theta = \sqrt{2}$, determine $\sec \theta$ using a Pythagorean identity.

Instructions: Derive the indicated equation by combining trig identities (and, where applicable, the definitions of secant, cosecant, and cotangent) together through substitution and applying algebra. There are no answers to these problems in the back of the book (since the answer is instead given in the question).

(1) Use trig identities to show that $\cot^2 \theta + \csc^2 \theta = 2 \csc^2 \theta - 1$.

(2) Use trig identities to show that $\tan^2 \theta + \sec^2 \theta = 1 + 2 \tan^2 \theta$.

(3) Use trig identities to show that $\sin^2 \theta - \cos^2 \theta = 1 - 2 \cos^2 \theta$.

(4) Use trig identities to show that $\cot^2 \theta + \csc^2 \theta = 1 + 2 \cot^2 \theta$.

(5) Use trig identities to show that $\sin^2 \theta - \cos^2 \theta = 2 \sin^2 \theta - 1$.

Instructions: Derive the indicated equation by combining trig identities (and, where applicable, the definitions of secant, cosecant, and cotangent) together through substitution and applying algebra. There are no answers to these problems in the back of the book (since the answer is instead given in the question).

(1) Use trig identities to show that $\sin\theta = \pm\frac{\tan\theta}{\sqrt{1+\tan^2\theta}}$.

(2) Use trig identities to show that $\tan\theta = \pm\frac{\sin\theta}{\sqrt{1-\sin^2\theta}}$.

(3) Use trig identities to show that $\sec\theta = \pm\frac{\sqrt{1+\cot^2\theta}}{\cot\theta}$.

Instructions: Derive the indicated equation by combining trig identities (and, where applicable, the definitions of secant, cosecant, and cotangent) together through substitution and applying algebra. There are no answers to these problems in the back of the book (since the answer is instead given in the question).

(1) Use trig identities to show that $\cos(3\theta) = 4\cos^3\theta - 3\cos\theta$.

(2) Use trig identities to show that $\tan(3\theta) = \frac{3\tan\theta - \tan^3\theta}{1 - 3\tan^2\theta}$.

(3) Use trig identities to show that $\cot(3\theta) = \frac{3\cot\theta - \cot^3\theta}{1 - 3\cot^2\theta}$.

Instructions: Derive the indicated equation by combining trig identities (and, where applicable, the definitions of secant, cosecant, and cotangent) together through substitution and applying algebra. There are no answers to these problems in the back of the book (since the answer is instead given in the question).

(1) Use trig identities to show that $\cos(2\theta) = \frac{1-\tan^2\theta}{1+\tan^2\theta}$.

(2) Use trig identities to show that $\sin(2\theta) = \frac{2\tan\theta}{1+\tan^2\theta}$.

(3) Use trig identities to show that $\cot(2\theta) = \frac{\cot^2\theta-1}{2\cot\theta}$.

Instructions: Derive the indicated equation by combining trig identities (and, where applicable, the definitions of secant, cosecant, and cotangent) together through substitution and applying algebra. There are no answers to these problems in the back of the book (since the answer is instead given in the question).

(1) Use trig identities to show that $\tan\left(\frac{\theta}{2}\right) = \frac{\sin\theta}{1+\cos\theta}$.

(2) Use trig identities to show that $\tan\left(\frac{\theta}{2}\right) = \csc\theta - \cot\theta$.

(3) Use trig identities to show that $\cot\left(\frac{\theta}{2}\right) = \csc\theta + \cot\theta$.

Chapter 11: Solve Algebraic Equations that Involve Trig Functions

In this chapter, we focus on how to solve algebraic equations that involve trig functions – in particular, how to solve for the angle. The basic algebraic steps are the same as those involved in solving for an unknown in any other algebraic equation: You generally collect the variables on one side of the equation and constants on the other side (though there is sometimes an exception to this rule, when there are two different trig functions present in the equation – as we will learn in an example), factor out the unknown (or apply the quadratic equation, if applicable), and divide by the coefficient of the unknown to solve for it. Since you're reading a trigonometry workbook, hopefully, you are already fluent in algebra. Therefore, we will focus on the trigonometry aspect of these algebra problems.

The main idea is this: Isolate the trig function, then apply the inverse of the trig function to both sides of the equation. When there are two different trig functions present, you must first apply one or more trig identities so that after substitutions only a single trig function remains. Sometimes, you may get a quadratic equation in the trig function. You solve this the same was that you would solve an ordinary quadratic equation, except that the result equals the trig function and then you apply the inverse trig function to both sides. These techniques are best illustrated by examples, so in this chapter we will keep the introduction very brief and move onto the examples.

Instructions: Determine the angles that solve each problem by applying algebra and, where applicable, trigonometric identities (see Chapter 10).

Procedure: Isolate the trig function, then apply the inverse of the trig function to both sides of the equation. When there are two different trig functions present, you must first apply one or more trig identities so that after substitutions only a single trig function remains. Sometimes, you may get a quadratic equation in the trig function. You solve this the same was that you would solve an ordinary quadratic equation, except that the result equals the trig function and then you apply the inverse trig function to both sides.

After isolating the trig function, follow the technique of Chapter 8 in order to find the angles corresponding to the inverse trig function. There may be a subtle point involved in this, especially if the argument of the trig function contains more than just the unknown by itself; see the examples for an illustration of this.

Example 1: $2 \sin \theta + 3 = 4$

First isolate the sine function, then take an inverse sine on both sides:
$$2 \sin \theta = 4 - 3 = 1$$
$$\sin \theta = 1/2$$
$$\theta = \sin^{-1}(1/2) = 30° \text{ or } 150°$$

Example 2: $3\tan(2\theta) = \sqrt{3}$

First isolate the tangent function, then take an inverse tangent on both sides:
$$\tan(2\theta) = \sqrt{3}/3$$
$$2\theta = \tan^{-1}(\sqrt{3}/3) = 30° \text{ or } 210°$$
We must set 2θ equal to each possible angle and solve for θ in each case: $2\theta = 30°$ leads to $\theta = 15°$, while $2\theta = 210°$ leads to $\theta = 105°$. The two solutions are $\theta = 15°$ and $\theta = 105°$.

Example 3: $2\sec\theta - 4 = 1 - 3\sec\theta$

First isolate the secant function, then take an inverse secant on both sides:
$$5\sec\theta = 5$$
$$\sec\theta = 1$$
$$\theta = \sec^{-1}1 = 0°$$

Example 4: $2\csc\theta = 4\sin\theta$

When there are two (or more) different trig functions, you must make one or more substitutions until a single trig function remains. In this case, we can write $\csc\theta = 1/\sin\theta$:
$$\frac{2}{\sin\theta} = 4\sin\theta$$
$$2 = 4\sin^2\theta$$
$$\sin^2\theta = \frac{1}{2}$$
$$\sin\theta = \pm\frac{1}{\sqrt{2}} = \pm\frac{1}{\sqrt{2}}\frac{\sqrt{2}}{\sqrt{2}} = \pm\frac{\sqrt{2}}{2}$$
$$\theta = \sin^{-1}\left(\pm\frac{\sqrt{2}}{2}\right) = 45°, 135°, 225°, \text{ or } 315°$$
There are four possible answers because there are two angles corresponding to each sign.

Example 5: $\sqrt{2}\sin\theta = -\sqrt{6}\cos\theta$

When there are two (or more) different trig functions, you must make one or more substitutions until a single trig function remains. In this case, we can use $\tan\theta = \sin\theta/\cos\theta$:
$$\sqrt{2}\tan\theta = -\sqrt{6}$$
$$\tan\theta = -\sqrt{3}$$
$$\theta = \tan^{-1}(-\sqrt{3}) = 120° \text{ or } 300°$$

Example 6: $5\cos\theta = 0$

First isolate the cosine function, then take an inverse cosine on both sides:
$$\cos\theta = 0$$
$$\theta = \cos^{-1}(0) = 90° \text{ or } 270°$$

Example 7: $2\sin(\theta/2) = -\sqrt{2}\tan(\theta/2)$

When there are two (or more) different trig functions, you must make one or more substitutions until a single trig function remains. In this case, we can use $\tan\theta = \sin\theta/\cos\theta$:

$$2\sin(\theta/2) = -\sqrt{2}\frac{\sin(\theta/2)}{\cos(\theta/2)}$$

It may be tempting to divide both sides by $\sin(\theta/2)$ here, but there are two good reasons not to do this: First, if $\sin(\theta/2)$ equals zero, you had better not divide by zero; and second, you will miss one possible solution. The correct approach is to factor $\sin(\theta/2)$ as follows:

$$2\sin(\theta/2) + \sqrt{2}\frac{\sin(\theta/2)}{\cos(\theta/2)} = 0$$

$$\sin(\theta/2)\left(2 + \sqrt{2}\frac{1}{\cos(\theta/2)}\right) = 0$$

$$\sin(\theta/2) = 0 \quad \text{or} \quad 2 + \sqrt{2}\frac{1}{\cos(\theta/2)} = 0$$

We must treat each case separately:

$$\frac{\theta}{2} = \sin^{-1}(0) \quad \text{or} \quad \sqrt{2}\frac{1}{\cos(\theta/2)} = -2$$

$$\frac{\theta}{2} = 0° \text{ or } 180° \quad \text{or} \quad -\frac{\sqrt{2}}{2} = \cos(\theta/2)$$

$$\theta = 0° \text{ or } 360° \quad \text{or} \quad \frac{\theta}{2} = \cos^{-1}\left(-\frac{\sqrt{2}}{2}\right) = 135° \text{ or } 225°$$

$$\theta = 0°, 270°, 360°, \text{ or } 450°$$

Example 8: $4\sin\theta\cos\theta = -\sqrt{3}$

When there are two (or more) different trig functions, you must make one or more substitutions until a single trig function remains. In this case, we can use the double-angle identity for sine, $\sin(2\theta) = 2\sin\theta\cos\theta$ (see Chapter 10):

$$2\sin(2\theta) = -\sqrt{3}$$
$$\sin(2\theta) = -\sqrt{3}/2$$
$$2\theta = \sin^{-1}(-\sqrt{3}/2) = 240° \text{ or } 300°$$
$$\theta = \sin^{-1}(-\sqrt{3}/2) = 120° \text{ or } 150°$$

Example 9: $5\sin^2\theta + \cos^2\theta = 2$

When there are two (or more) different trig functions, you must make one or more substitutions until a single trig function remains. In this case, we can use the Pythagorean identity, $\sin^2\theta + \cos^2\theta = 1$ (see Chapter 10):

$$5\sin^2\theta + \cos^2\theta = (4\sin^2\theta + \sin^2\theta) + \cos^2\theta = 4\sin^2\theta + 1 = 2$$
$$4\sin^2\theta = 1$$
$$\sin\theta = \pm1/2$$
$$\theta = \sin^{-1}(\pm1/2) = 30°, 150°, 210°, \text{ or } 330°$$

Example 10: $2\cos^2\theta = 3 + \cos\theta$

This is the quadratic equation – i.e. it is quadratic in $\cos\theta$. If you define x to be $\cos\theta$, the equation looks like $2x^2 = 3 + x$. The first step toward solving the quadratic equation is to cast the equation in standard form, which is $ax^2 + bx + c = 0$. To do this, we move the terms around (following the rules of algebra, of course):

$$2x^2 - x - 3 = 0$$

We can now identify the constants of the quadratic equation:

$$a = 2 \quad , \quad b = -1 \quad , \quad c = -3$$

The solution to the quadratic equation, $ax^2 + bx + c = 0$, is:

$$x = \frac{-b \pm \sqrt{b^2 - 4ac}}{2a} = \frac{-(-1) \pm \sqrt{(-1)^2 - 4(2)(-3)}}{2(2)} = \frac{1 \pm \sqrt{1 + 24}}{4} = \frac{1 \pm \sqrt{25}}{4} = \frac{1 \pm 5}{4}$$

$$x = \frac{1 - 5}{4} \text{ or } \frac{1 + 5}{4}$$

$$x = -1 \text{ or } 3/2$$

Recall that $x = \cos\theta$:

$$\theta = \cos^{-1}(-1) \text{ or } \cos^{-1}(3/2)$$

The second case does not have a real solution, so the only answer is $\theta = 180°$.

Example 11: $\sqrt{3}\csc\theta - \sqrt{3} = \cot\theta$

When there are two (or more) different trig functions, you must make one or more substitutions until a single trig function remains. In this case, we can use the Pythagorean identity, $1 + \cot^2\theta = \csc^2\theta$ (see Chapter 10), in the form $\cot\theta = \pm\sqrt{\csc^2\theta - 1}$:

$$\sqrt{3}\csc\theta - \sqrt{3} = \pm\sqrt{\csc^2\theta - 1}$$

Eliminate the squareroot by first isolating it (it already is in this problem), and then squaring both sides:

$$\left(\sqrt{3}\csc\theta - \sqrt{3}\right)^2 = \csc^2\theta - 1$$

$$3\csc^2\theta - 6\csc\theta + 3 = \csc^2\theta - 1$$

$$2\csc^2\theta - 6\csc\theta + 4 = 0$$

This is the quadratic equation – i.e. it is quadratic in $\csc\theta$. If you define x to be $\csc\theta$, the equation looks like $2x^2 - 6x + 4 = 0$. We solve this just like the previous example. First, we identify the constants:

$$a = 2 \quad , \quad b = -6 \quad , \quad c = 4$$

The solution to the quadratic equation, $ax^2 + bx + c = 0$, is:

$$x = \frac{-b \pm \sqrt{b^2 - 4ac}}{2a} = \frac{-(-6) \pm \sqrt{(-6)^2 - 4(2)(4)}}{2(2)} = \frac{6 \pm \sqrt{36 - 32}}{4} = \frac{6 \pm \sqrt{4}}{4} = \frac{6 \pm 2}{4}$$

$$x = \frac{6 - 2}{4} \text{ or } \frac{6 + 2}{4}$$

$$x = 1 \text{ or } 2$$

Recall that $x = \csc\theta$:

$$\theta = \csc^{-1}1 \text{ or } \csc^{-1}(2) = 30° \text{ or } 90°$$

The alternate angle, $150°$, does not satisfy the original equation.

168

Instructions: Determine the angles that solve each problem by applying algebra and, where applicable, trigonometric identities. Check your answers in the back of the book.

(1) $2 \sin \theta = -\sqrt{3}$

(2) $2 = \sqrt{2} \sec \theta$

(3) $3 \tan \theta - 2 = \tan \theta - 4$

(4) $5 \cos \theta - 2\sqrt{3} = \sqrt{3} - \cos \theta$

(5) $4 \csc(2\theta) = -8$

(6) $4 \cot^2 \theta - 2 = 2$

Instructions: Determine the angles that solve each problem by applying algebra and, where applicable, trigonometric identities. Check your answers in the back of the book.

(1) $\tan(2\theta) - \sqrt{3} = 2\tan(2\theta)$

(2) $\sqrt{3}\csc(\theta/2) = -2$

(3) $\sqrt{3} = -3\cot(3\theta)$

(4) $8\sin^2(2\theta) = 2$

(5) $\sqrt{6}\cos(\theta + \pi/4) = \sqrt{3}$

(6) $3\sec(3\theta) = \sec(3\theta) - 4$

Instructions: Determine the angles that solve each problem by applying algebra and, where applicable, trigonometric identities. Check your answers in the back of the book.

(1) $\sqrt{2}\cos\theta - \sin^2\theta = \cos^2\theta$

(2) $\sin^2\theta + 3\cos^2\theta = 3$

(3) $1 - \sec^2(2\theta) = \tan^2(2\theta)$

(4) $2\cos^2\theta - 2\sin^2\theta = -\sqrt{3}$

(5) $\sqrt{2} = -4\sin\theta\cos\theta$

(6) $\sin^2\theta - 1 = \cos^2\theta$

Instructions: Determine the angles that solve each problem by applying algebra and, where applicable, trigonometric identities. Check your answers in the back of the book.

(1) $4\cos\theta = 3\sec\theta$

(2) $\tan(2\theta) = 3\cot(2\theta)$

(3) $\tan\theta = -2\sin\theta$

(4) $3\csc\theta = -6\cot\theta$

(5) $\sin(2\theta) = \cos\theta$

(6) $2\tan\theta = -\sqrt{2}\sec\theta$

Instructions: Determine the angles that solve each problem by applying algebra and, where applicable, trigonometric identities. Check your answers in the back of the book.

(1) $2\sin^2\theta = 1 + \sin\theta$

(2) $3 + 4\cos^2\theta = 4\sqrt{3}\cos\theta$

(3) $2 - \csc^2\theta = \csc\theta$

(4) $2\sqrt{3}\tan(2\theta) + \tan^2(2\theta) = -3$

Instructions: Determine the angles that solve each problem by applying algebra and, where applicable, trigonometric identities. Check your answers in the back of the book.

(1) $2\sqrt{2}\sec\theta - 2 = \sec^2\theta$

(2) $2\sin^2(3\theta) = 3\sin(3\theta) - 1$

(3) $\sqrt{3}\cos\theta - \sqrt{3} = \sin\theta$

(4) $2\cos\theta = 3\tan\theta$

Answer Key

Chapter 1 Answers:

Page 7
(1) π/5 rad (2) π/10 rad
(3) 13π/3 rad (4) 13π/20 rad
(5) π/4 rad (6) π/15 rad
(7) 8π/15 rad (8) 3π/2 rad
(9) 2π/3 rad (10) 13π/30 rad
(11) 4π/15 rad (12) 5π/6 rad
Page 8
(1) 7π/6 rad (2) π/4 rad
(3) 2π/15 rad (4) π/60 rad
(5) π/18 rad (6) 3π/4 rad
(7) 8π/15 rad (8) 7π/4 rad
(9) 3π/2 rad (10) 7π/30 rad
(11) 29π/60 rad (12) 23π/60 rad
Page 9
(1) 17π/20 rad (2) 4π/15 rad
(3) 19π/60 rad (4) 7π/20 rad
(5) 8π/15 rad (6) 46π/15 rad
(7) 101π/30 rad (8) 11π/15 rad
(9) 5π/4 rad (10) 5π/3 rad
(11) 43π/60 rad (12) π/3 rad
Page 10
(1) 29π/15 rad (2) 13π/20 rad
(3) 7π/12 rad (4) 68π/15 rad
(5) π/2 rad (6) 19π/60 rad
(7) 3π/20 rad (8) 23π/60 rad
(9) 35π/12 rad (10) 23π/15 rad
(11) 7π/20 rad (12) 53π/15 rad
Page 11
(1) 23π/30 rad (2) 13π/60 rad
(3) 229π/60 rad (4) 91π/60 rad
(5) 17π/60 rad (6) 2π/5 rad
(7) 11π/30 rad (8) 8π/5 rad
(9) π/6 rad (10) 2π/3 rad
(11) 11π/20 rad (12) 4π/5 rad

Page 12
(1) 17π/6 rad (2) 151π/60 rad
(3) 101π/60 rad (4) 53π/60 rad
(5) 169π/60 rad (6) 47π/60 rad
(7) 11π/20 rad (8) 19π/30 rad
(9) 17π/30 rad (10) 46π/15 rad
(11) 3π rad (12) 23π/30 rad
Page 13
(1) 37π/15 rad (2) 271π/60 rad
(3) 3π/10 rad (4) 13π/20 rad
(5) 7π/20 rad (6) 29π/12 rad
(7) 3π/20 rad (8) 31π/60 rad
(9) 61π/15 rad (10) 17π/60 rad
(11) 7π/10 rad (12) 2π/5 rad
Page 14
(1) 5π/12 rad (2) 59π/60 rad
(3) 59π/15 rad (4) 14π/15 rad
(5) π/20 rad (6) 91π/20 rad
(7) 31π/60 rad (8) 3π/5 rad
(9) π rad (10) 13π/60 rad
(11) 61π/30 rad (12) π/60 rad
Page 15
(1) 11π/20 rad (2) 8π/15 rad
(3) 28π/15 rad (4) 9π/5 rad
(5) 19π/30 rad (6) 33π/20 rad
(7) 13π/15 rad (8) 2π rad
(9) 5π/3 rad (10) 49π/60 rad
(11) π/2 rad (12) π/3 rad
Page 16
(1) 9π/10 rad (2) 9π/20 rad
(3) 7π/20 rad (4) 17π/60 rad
(5) 11π/30 rad (6) π/4 rad
(7) 5π/12 rad (8) 7π/12 rad
(9) 13π/30 rad (10) 29π/60 rad
(11) 3π/10 rad (12) 19π/60 rad

Page 17
(1) 37π/60 rad (2) 53π/60 rad
(3) 29π/60 rad (4) 77π/60 rad
(5) 7π/15 rad (6) π/20 rad
(7) 28π/15 rad (8) π/6 rad
(9) 37π/20 rad (10) 19π/20 rad
(11) 4π/15 rad (12) π/30 rad
Page 18
(1) 3π/5 rad (2) 17π/20 rad
(3) 7π/30 rad (4) 7π/10 rad
(5) 17π/10 rad (6) 13π/20 rad
(7) π/2 rad (8) 5π/3 rad
(9) 73π/60 rad (10) 31π/60 rad
(11) 7π/20 rad (12) 2π/5 rad

Chapter 2 Answers:

Page 20
(1) 81° (2) 9°
(3) 102° (4) 63°
(5) 174° (6) 15°
(7) 156° (8) 423°
(9) 180° (10) 45°
(11) 834° (12) 90°
Page 21
(1) 651° (2) 75°
(3) 153° (4) 114°
(5) 120° (6) 588°
(7) 321° (8) 363°
(9) 102° (10) 21°
(11) 36° (12) 63°
Page 22
(1) 3° (2) 102°
(3) 9° (4) 810°
(5) 90° (6) 129°
(7) 57° (8) 144°
(9) 54° (10) 474°
(11) 96° (12) 165°

Page 23
(1) 66° (2) 45°
(3) 96° (4) 87°
(5) 57° (6) 99°
(7) 351° (8) 24°
(9) 15° (10) 162°
(11) 81° (12) 54°
Page 24
(1) 288° (2) 39°
(3) 828° (4) 558°
(5) 369° (6) 168°
(7) 12° (8) 81°
(9) 141° (10) 147°
(11) 174° (12) 87°
Page 25
(1) 63° (2) 714°
(3) 39° (4) 78°
(5) 111° (6) 75°
(7) 141° (8) 123°
(9) 27° (10) 156°
(11) 114° (12) 117°
Page 26
(1) 681° (2) 162°
(3) 141° (4) 27°
(5) 132° (6) 6°
(7) 678° (8) 117°
(9) 24° (10) 156°
(11) 426° (12) 21°
Page 27
(1) 279° (2) 180°
(3) 93° (4) 195°
(5) 540° (6) 150°
(7) 147° (8) 474°
(9) 54° (10) 111°
(11) 144° (12) 87°
Page 28
(1) 735° (2) 114°
(3) 27° (4) 96°
(5) 12° (6) 480°
(7) 48° (8) 66°
(9) 120° (10) 54°
(11) 108° (12) 744°

Page 29
(1) 105° (2) 378°
(3) 210° (4) 21°
(5) 621° (6) 651°
(7) 150° (8) 681°
(9) 705° (10) 30°
(11) 63° (12) 81°
Page 30
(1) 132° (2) 165°
(3) 162° (4) 534°
(5) 57° (6) 60°
(7) 123° (8) 450°
(9) 96° (10) 246°
(11) 141° (12) 84°
Page 31
(1) 870° (2) 165°
(3) 489° (4) 21°
(5) 522° (6) 96°
(7) 60° (8) 111°
(9) 141° (10) 57°
(11) 126° (12) 48°

Chapter 3 Answers:

Page 35
(1) 4/5, 3/5 (2) 5/13, 5/12
(3) $2\sqrt{5}/5$, 1/2 (4) $\sqrt{3}/2$, $\sqrt{3}/3$
(5) 4/5, 5/3 (6) $\sqrt{2}$, $\sqrt{2}/2$
Page 36
(1) $3\sqrt{13}/13$, 2/3 (2) $\sqrt{2}/2$, $\sqrt{2}/2$
(3) 1/2, $\sqrt{3}$ (4) $2\sqrt{5}/5$, 2
(5) $\sqrt{6}/2$, $\sqrt{3}$ (6) 5/4, 4/3
Page 37
(1) 3/5, 4/5 (2) $\sqrt{22}/11$, $\sqrt{2}/3$
(3) $\sqrt{2}/2$, 1 (4) $\sqrt{6}/3$, $\sqrt{2}/2$
(5) 3/5, 5/4 (6) $2\sqrt{3}/3$, 1/2
Page 38
(1) 8/17, 15/8 (2) $7\sqrt{2}/10$, 1/7
(3) 4/5, 5/3 (4) $2\sqrt{13}/13$, 3/2
(5) $2\sqrt{5}/5$, $\sqrt{5}/5$ (6) $\sqrt{10}/3$, 1/3

Page 39
(1) 12/13, 5/13 (2) $\sqrt{2}$, 1
(3) $\sqrt{3}$, $\sqrt{2}/2$ (4) 3/5, 3/4
(5) $3\sqrt{13}/13$, $\sqrt{13}/2$ (6) $\sqrt{5}/5$, 1/2
Page 40
(1) 17/15, 17/8 (2) $\sqrt{2}$, 1
(3) 5/3, 4/3 (4) $\sqrt{5}/5$, 2
(5) 2/3, $3\sqrt{5}/5$ (6) 2, $\sqrt{3}/3$
Page 41
(1) $3\sqrt{10}/10$, 3 (2) $\sqrt{5}/3$, 2/3
(3) $\sqrt{21}/3$, $2\sqrt{3}/3$ (4) $\sqrt{3}/2$, $\sqrt{3}$
(5) $2\sqrt{2}/3$, 3 (6) 3/2, $\sqrt{5}/3$
Page 42
(1) $\sqrt{2}/2$, 1 (2) 4/5, 4/3
(3) $\sqrt{3}/2$, $\sqrt{3}$ (4) $\sqrt{2}$, 1
(5) $2\sqrt{3}/3$, 2 (6) 13/5, 12/5
Page 43
(1) 7/3, $2\sqrt{10}/7$ (2) $\sqrt{2}/2$, $\sqrt{2}$
(3) $2\sqrt{5}/5$, 1/2 (4) 3/5, 4/3
(5) 1/2, $2\sqrt{3}/3$ (6) 1/2, $\sqrt{3}/2$
Page 44
(1) 4/5, 4/3 (2) 12/13, 12/5
(3) $\sqrt{2}/2$, $\sqrt{2}$ (4) $2\sqrt{5}/5$, 2
(5) 1/2, $\sqrt{3}/2$ (6) $\sqrt{6}/2$, $\sqrt{2}$
Page 45
(1) $2\sqrt{6}/7$, 5/7 (2) $\sqrt{3}$, $\sqrt{2}$
(3) $3\sqrt{5}/5$, $\sqrt{5}/2$ (4) $\sqrt{15}/4$, $\sqrt{15}$
(5) 3/5, 5/4 (6) $\sqrt{3}/2$, $\sqrt{3}$
Page 46
(1) $\sqrt{5}/2$, $\sqrt{5}$ (2) $11\sqrt{10}/20$, $2\sqrt{10}/9$
(3) 17/15, 8/15 (4) $\sqrt{5}/5$, 2
(5) 2/3, $3\sqrt{5}/5$ (6) 12/13, 5/13

Chapter 4 Answers:

Page 51
(1) 12, $6\sqrt{3}$ (2) 45°, $4\sqrt{2}$
(3) $4\sqrt{3}$, 4 (4) 6, 6
(5) 30°, 4 (6) 20, 10
Page 52
(1) 60°, $14\sqrt{3}$ (2) $3\sqrt{2}$, 3
(3) 6, $3\sqrt{3}$ (4) 30°, $9\sqrt{3}$
(5) 45°, 5 (6) 6, $6\sqrt{3}$
Page 53
(1) $2\sqrt{2}$, 4 (2) $3\sqrt{3}$, 9
(3) $16\sqrt{3}/3$, $8\sqrt{3}/3$ (4) $\sqrt{6}$, $2\sqrt{2}$
(5) 30°, $2\sqrt{2}$ (6) $\sqrt{6}$, $\sqrt{3}$
Page 54
(1) $3\sqrt{6}$, $3\sqrt{2}$ (2) $\sqrt{6}$, $2\sqrt{3}$
(3) 30°, $\sqrt{2}$ (4) 1, $\sqrt{3}/2$
(5) 1/2, $\sqrt{2}/2$ (6) $\sqrt{2}$, $\sqrt{2}/2$
Page 55
(1) 45°, $\sqrt{5}$ (2) 60°, $\sqrt{7}$
(3) $8\sqrt{6}$, $4\sqrt{6}$ (4) $4\sqrt{6}$, $4\sqrt{6}$
(5) 60°, $\sqrt{15}$ (6) 30°, 14
Page 56
(1) 30°, $2\sqrt{5}$ (2) 45°, $\sqrt{7}$
(3) 32, $16\sqrt{3}$ (4) 60°, $4\sqrt{3}$
(5) 30°, $12\sqrt{3}$ (6) $8\sqrt{3}/3$, $16\sqrt{3}/3$
Page 57
(1) $12\sqrt{3}$, 18 (2) 60°, $8\sqrt{3}$
(3) 12, $12\sqrt{3}$ (4) 10, 10
(5) 30°, $4\sqrt{3}$ (6) 45°, 16
Page 58
(1) 60°, 21 (2) 18, $9\sqrt{2}$
(3) 40, 20 (4) 60°, 8
(5) 60°, 5 (6) $24\sqrt{2}$, $24\sqrt{2}$
Page 59
(1) $2\sqrt{14}$, $4\sqrt{7}$ (2) $3\sqrt{21}$, $9\sqrt{7}$
(3) 18, 9 (4) $\sqrt{11}$, $2\sqrt{11}$
(5) 30°, $2\sqrt{13}$ (6) $2\sqrt{11}$, $\sqrt{22}$

Page 60
(1) $18\sqrt{3}$, 18 (2) 5, 5
(3) 60°, $18\sqrt{2}$ (4) $2\sqrt{3}/3$, 1
(5) $\sqrt{2}/2$, 1 (6) $\sqrt{2}/3$, $\sqrt{2}/6$
Page 61
(1) 30°, $2\sqrt{17}$ (2) 45°, $\sqrt{19}$
(3) $2\sqrt{14}$, $\sqrt{14}$ (4) 45, $15\sqrt{3}$
(5) 45°, $\sqrt{10}$ (6) 30°, $12\sqrt{2}$
Page 62
(1) 45°, $2\sqrt{5}$ (2) 30°, $\sqrt{5}$
(3) $2\sqrt{30}$, $3\sqrt{10}$ (4) 60°, $5\sqrt{30}$
(5) 30°, $45\sqrt{10}$ (6) $70\sqrt{210}/3$, $140\sqrt{210}/3$

Chapter 5 Answers:

Page 67
(1) $2\sqrt{3}/3$ (2) 1/2 (3) $\sqrt{2}/2$
(4) undefined (5) 1/2 (6) $\sqrt{3}$
(7) 0 (8) 1/2 (9) 0
(10) $\sqrt{2}$ (11) $\sqrt{3}/3$ (12) undefined
(13) undefined (14) $2\sqrt{3}/3$ (15) 2
(16) $\sqrt{3}/2$ (17) $\sqrt{2}$ (18) 1/2
(19) $\sqrt{2}/2$ (20) 1 (21) 1
(22) 2 (23) 0 (24) 1
Page 68
(1) $\sqrt{2}$ (2) 2 (3) 2
(4) 1 (5) $\sqrt{2}$ (6) $\sqrt{2}$
(7) undefined (8) $\sqrt{2}$ (9) 1/2
(10) undefined (11) 0 (12) undefined
(13) 1 (14) 1 (15) 1
(16) 1/2 (17) undefined (18) 1/2
(19) 0 (20) undefined (21) 1/2
(22) 1/2 (23) $\sqrt{2}/2$ (24) 0

Page 69
(1) 1 (2) undefined (3) 1
(4) 1 (5) $2\sqrt{3}/3$ (6) undefined
(7) $\sqrt{2}$ (8) 0 (9) undefined
(10) $\sqrt{3}/2$ (11) 2 (12) $\sqrt{3}/2$
(13) 1/2 (14) undefined (15) $\sqrt{3}$
(16) $\sqrt{3}/3$ (17) $2\sqrt{3}/3$ (18) 2
(19) 1 (20) undefined (21) 1
(22) $2\sqrt{3}/3$ (23) 0 (24) $2\sqrt{3}/3$
Page 70
(1) 1/2 (2) 1 (3) $\sqrt{3}/3$
(4) 1/2 (5) 1 (6) 0
(7) 1 (8) $2\sqrt{3}/3$ (9) 2
(10) 2 (11) 1 (12) $\sqrt{2}/2$
(13) 0 (14) undefined (15) $\sqrt{2}$
(16) 0 (17) undefined (18) 1
(19) 0 (20) $\sqrt{3}/3$ (21) 1
(22) 1 (23) $\sqrt{3}/2$ (24) $\sqrt{3}$
Page 71
(1) undefined (2) 1 (3) 0
(4) 0 (5) undefined (6) 1
(7) undefined (8) undefined (9) 1
(10) undefined (11) $2\sqrt{3}/3$ (12) 1/2
(13) undefined (14) $\sqrt{3}$ (15) $2\sqrt{3}/3$
(16) 1 (17) undefined (18) $2\sqrt{3}/3$
(19) 1 (20) 0 (21) 2
(22) 1 (23) $\sqrt{3}/2$ (24) 0
Page 72
(1) 1 (2) $\sqrt{3}$ (3) 0
(4) 1/2 (5) 1/2 (6) $\sqrt{2}$
(7) $\sqrt{3}/3$ (8) $\sqrt{2}$ (9) $\sqrt{3}/3$
(10) $\sqrt{2}/2$ (11) 1 (12) $\sqrt{2}$
(13) $\sqrt{3}$ (14) $\sqrt{3}$ (15) $\sqrt{3}$
(16) 1/2 (17) undefined (18) 2
(19) $2\sqrt{3}/3$ (20) undefined (21) 0
(22) 0 (23) 0 (24) $\sqrt{3}/2$

Page 73
(1) $2\sqrt{3}/3$ (2) $\sqrt{2}/2$ (3) undefined
(4) 1 (5) $\sqrt{2}/2$ (6) $\sqrt{2}/2$
(7) $\sqrt{2}$ (8) $\sqrt{2}/2$ (9) 1
(10) 1 (11) $\sqrt{3}/3$ (12) 1
(13) 0 (14) $2\sqrt{3}/3$ (15) 2
(16) 1 (17) undefined (18) 0
(19) undefined (20) $\sqrt{2}$ (21) 1
(22) $\sqrt{2}$ (23) undefined (24) 2
Page 74
(1) 2 (2) undefined (3) $\sqrt{2}/2$
(4) undefined (5) $\sqrt{2}$ (6) 1
(7) $\sqrt{3}/2$ (8) $\sqrt{3}/3$ (9) 1/2
(10) undefined (11) 1 (12) $2\sqrt{3}/3$
(13) 1 (14) 1/2 (15) $\sqrt{3}$
(16) 2 (17) $\sqrt{3}$ (18) $\sqrt{2}$
(19) undefined (20) 1 (21) 1
(22) $\sqrt{2}/2$ (23) 0 (24) 0
Page 75
(1) 1/2 (2) $\sqrt{3}/2$ (3) $\sqrt{3}$
(4) undefined (5) 1 (6) undefined
(7) undefined (8) $\sqrt{2}$ (9) 1/2
(10) undefined (11) $\sqrt{2}$ (12) 1
(13) undefined (14) 1 (15) $\sqrt{2}$
(16) 1 (17) undefined (18) 2
(19) $\sqrt{3}$ (20) $\sqrt{3}/3$ (21) 0
(22) 1 (23) $2\sqrt{3}/3$ (24) $\sqrt{3}/2$
Page 76
(1) $\sqrt{3}/3$ (2) $\sqrt{3}/3$ (3) 1
(4) 1/2 (5) 1 (6) 0
(7) 1 (8) 0 (9) 0
(10) 1/2 (11) 2 (12) undefined
(13) 1 (14) $\sqrt{3}/2$ (15) undefined
(16) 0 (17) 1 (18) $\sqrt{3}$
(19) 1 (20) $\sqrt{3}/3$ (21) $\sqrt{2}$
(22) 1 (23) undefined (24) 1

Page 77
(1) $\sqrt{2}/2$ (2) 1/2 (3) undefined
(4) 1/2 (5) undefined (6) 1
(7) $\sqrt{3}/3$ (8) 1/2 (9) $\sqrt{3}/3$
(10) undefined (11) undefined (12) 1
(13) undefined (14) 1 (15) $\sqrt{3}/2$
(16) $\sqrt{3}$ (17) 1 (18) $\sqrt{3}/2$
(19) 1 (20) $\sqrt{3}/3$ (21) $\sqrt{3}/3$
(22) $\sqrt{2}/2$ (23) 1 (24) 0
Page 78
(1) 2 (2) 2 (3) 2
(4) undefined (5) 1 (6) 2
(7) $\sqrt{2}/2$ (8) 1 (9) $2\sqrt{3}/3$
(10) 1/2 (11) $\sqrt{2}$ (12) 2
(13) 0 (14) 0 (15) $\sqrt{2}$
(16) 2 (17) $2\sqrt{3}/3$ (18) $\sqrt{3}/2$
(19) $\sqrt{2}/2$ (20) $2\sqrt{3}/3$ (21) $2\sqrt{3}/3$
(22) 0 (23) $\sqrt{3}$ (24) $\sqrt{3}/2$

Chapter 6 Answers:

Page 82
(1) 23° (2) 29° (3) 8°
(4) 15° (5) 19° (6) 82°
(7) 71° (8) 28° (9) 45°
(10) 64° (11) 40° (12) 1°
(13) 76° (14) 69° (15) 45°
(16) 69° (17) 72° (18) 28°
Page 83
(1) 24° (2) 11° (3) 20°
(4) 33° (5) 31° (6) 88°
(7) 35° (8) 15° (9) 50°
(10) 71° (11) 36° (12) 14°
(13) 79° (14) 84° (15) 63°
(16) 70° (17) 4° (18) 52°

Page 84
(1) 49° (2) 22° (3) 20°
(4) 82° (5) 48° (6) 35°
(7) 81° (8) 15° (9) 60°
(10) 49° (11) 52° (12) 29°
(13) 40° (14) 34° (15) 28°
(16) 30° (17) 82° (18) 15°
Page 85
(1) 47° (2) 58° (3) 19°
(4) 7° (5) 21° (6) 50°
(7) 17° (8) 48° (9) 44°
(10) 4° (11) 26° (12) 77°
(13) 40° (14) 34° (15) 15°
(16) 51° (17) 15° (18) 60°
Page 86
(1) 73° (2) 20° (3) 56°
(4) 60° (5) 16° (6) 28°
(7) 45° (8) 84° (9) 4°
(10) 51° (11) 64° (12) 35°
(13) 38° (14) 5° (15) 40°
(16) 13° (17) 58° (18) 34°
Page 87
(1) 11° (2) 52° (3) 60°
(4) 66° (5) 38° (6) 2°
(7) 70° (8) 40° (9) 25°
(10) 9° (11) 64° (12) 16°
(13) 76° (14) 11° (15) 23°
(16) 55° (17) 60° (18) 88°
Page 88
(1) 29° (2) 38° (3) 54°
(4) 29° (5) 32° (6) 50°
(7) 30° (8) 34° (9) 7°
(10) 79° (11) 86° (12) 85°
(13) 63° (14) 3° (15) 1°
(16) 5° (17) 48° (18) 43°
Page 89
(1) 77° (2) 27° (3) 22°
(4) 46° (5) 58° (6) 23°
(7) 27° (8) 63° (9) 72°
(10) 15° (11) 11° (12) 26°
(13) 61° (14) 57° (15) 15°
(16) 29° (17) 84° (18) 25°

Page 90
(1) 40° (2) 72° (3) 29°
(4) 17° (5) 76° (6) 51°
(7) 29° (8) 58° (9) 59°
(10) 50° (11) 5° (12) 50°
(13) 13° (14) 50° (15) 45°
(16) 84° (17) 77° (18) 26°
Page 91
(1) 15° (2) 42° (3) 30°
(4) 1° (5) 11° (6) 49°
(7) 59° (8) 29° (9) 49°
(10) 62° (11) 53° (12) 24°
(13) 3° (14) 10° (15) 35°
(16) 83° (17) 86° (18) 1°
Page 92
(1) 39° (2) 81° (3) 49°
(4) 2° (5) 83° (6) 77°
(7) 71° (8) 25° (9) 50°
(10) 85° (11) 67° (12) 16°
(13) 71° (14) 23° (15) 74°
(16) 47° (17) 62° (18) 27°
Page 93
(1) 59° (2) 35° (3) 17°
(4) 35° (5) 42° (6) 18°
(7) 30° (8) 55° (9) 89°
(10) 77° (11) 54° (12) 54°
(13) 34° (14) 64° (15) 78°
(16) 69° (17) 52° (18) 71°

Chapter 7 Answers:

Page 97
(1) 2 (2) $-\sqrt{2}$ (3) -1
(4) -1 (5) $-1/2$ (6) $-2\sqrt{3}/3$
(7) $-\sqrt{2}/2$ (8) $1/2$ (9) $-\sqrt{3}/2$
(10) $\sqrt{2}$ (11) $-\sqrt{3}/3$ (12) -2
(13) $-\sqrt{3}$ (14) 1 (15) $-\sqrt{3}/2$
(16) -1 (17) undefined (18) $-\sqrt{2}$
(19) $-2\sqrt{3}/3$ (20) $-\sqrt{2}$ (21) -1
(22) -1 (23) $-\sqrt{3}$ (24) 1

Page 98
(1) $-\sqrt{3}$ (2) -1 (3) -1
(4) 1 (5) $2\sqrt{3}/3$ (6) $-\sqrt{2}/2$
(7) $-1/2$ (8) $1/2$ (9) $-\sqrt{3}/3$
(10) -1 (11) undefined (12) 1
(13) $-\sqrt{3}/2$ (14) $-2\sqrt{3}/3$ (15) -2
(16) -1 (17) $-2\sqrt{3}/3$ (18) $-\sqrt{3}/2$
(19) undefined (20) $-2\sqrt{3}/3$ (21) 2
(22) 0 (23) -1 (24) -1
Page 99
(1) undefined (2) $-\sqrt{3}/3$ (3) $1/2$
(4) -1 (5) $-\sqrt{2}$ (6) $-2\sqrt{3}/3$
(7) $-\sqrt{3}/2$ (8) $\sqrt{3}/2$ (9) $1/2$
(10) $-\sqrt{3}/3$ (11) -2 (12) $-\sqrt{3}/3$
(13) $-1/2$ (14) $-\sqrt{3}/2$ (15) $-\sqrt{2}$
(16) $-\sqrt{3}$ (17) $-1/2$ (18) undefined
(19) undefined (20) $-\sqrt{3}/2$ (21) $\sqrt{2}/2$
(22) $-\sqrt{2}$ (23) -1 (24) $-\sqrt{3}/2$
Page 100
(1) undefined (2) 0 (3) $-\sqrt{3}/3$
(4) -2 (5) $\sqrt{3}/3$ (6) $-\sqrt{3}/3$
(7) -2 (8) undefined (9) 1
(10) undefined (11) $-\sqrt{2}$ (12) $1/2$
(13) 2 (14) -1 (15) -2
(16) $-1/2$ (17) $\sqrt{2}$ (18) 0
(19) 0 (20) undefined (21) -1
(22) $-\sqrt{2}/2$ (23) $1/2$ (24) -1
Page 101
(1) 1 (2) $-\sqrt{3}/2$ (3) $-\sqrt{3}/3$
(4) 1 (5) $\sqrt{3}$ (6) 0
(7) $\sqrt{2}/2$ (8) $\sqrt{2}/2$ (9) $-\sqrt{3}/3$
(10) 0 (11) -2 (12) 1
(13) 1 (14) $\sqrt{2}/2$ (15) $\sqrt{3}/2$
(16) $-\sqrt{2}$ (17) 1 (18) -1
(19) $1/2$ (20) $\sqrt{2}$ (21) 2
(22) $-\sqrt{2}$ (23) 1 (24) $-2\sqrt{3}/3$

Page 102
(1) $-\sqrt{3}/3$ (2) -1 (3) $-\sqrt{2}$
(4) 1 (5) $-\sqrt{3}/3$ (6) 0
(7) 2 (8) $-\sqrt{2}/2$ (9) $-1/2$
(10) $\sqrt{3}$ (11) 1 (12) 1
(13) 2 (14) 1 (15) $-\sqrt{3}/2$
(16) $-\sqrt{3}/2$ (17) $-1/2$ (18) $2\sqrt{3}/3$
(19) -1 (20) -1 (21) $-2\sqrt{3}/3$
(22) -1 (23) $-2\sqrt{3}/3$ (24) $-1/2$

Page 103
(1) $2\sqrt{3}/3$ (2) -2 (3) -1
(4) $-\sqrt{3}/2$ (5) $-\sqrt{3}/2$ (6) undefined
(7) 1 (8) 1 (9) -1
(10) $-\sqrt{3}/3$ (11) 2 (12) undefined
(13) $-\sqrt{2}$ (14) $\sqrt{3}/2$ (15) -1
(16) $1/2$ (17) $-1/2$ (18) $-\sqrt{3}$
(19) 1 (20) -2 (21) $\sqrt{3}/3$
(22) $\sqrt{2}/2$ (23) $1/2$ (24) $-\sqrt{3}$

Page 104
(1) $-2\sqrt{3}/3$ (2) $\sqrt{3}$ (3) $-\sqrt{2}$
(4) $-\sqrt{3}/2$ (5) $\sqrt{2}$ (6) $-\sqrt{3}$
(7) 1 (8) undefined (9) 1
(10) -1 (11) 0 (12) $-\sqrt{2}/2$
(13) $-1/2$ (14) $-\sqrt{3}$ (15) $-\sqrt{3}$
(16) $1/2$ (17) $-\sqrt{3}$ (18) -2
(19) -1 (20) -2 (21) -2
(22) $\sqrt{3}/3$ (23) $-\sqrt{2}$ (24) undefined

Page 105
(1) undefined (2) $\sqrt{3}/3$ (3) undefined
(4) $-\sqrt{3}/3$ (5) $2\sqrt{3}/3$ (6) $\sqrt{3}/2$
(7) $-\sqrt{2}/2$ (8) -1 (9) -1
(10) -1 (11) $-\sqrt{2}/2$ (12) $-1/2$
(13) $\sqrt{3}/3$ (14) $-\sqrt{3}/2$ (15) -2
(16) $-\sqrt{3}/2$ (17) $\sqrt{2}/2$ (18) $-\sqrt{2}/2$
(19) undefined (20) $\sqrt{3}/2$ (21) 0
(22) undefined (23) undefined (24) 1

Page 106
(1) 0 (2) $\sqrt{3}/3$ (3) 0
(4) -2 (5) $-\sqrt{2}/2$ (6) undefined
(7) $-2\sqrt{3}/3$ (8) undefined (9) -1
(10) -2 (11) $\sqrt{3}/2$ (12) -1
(13) $-\sqrt{3}$ (14) undefined (15) $-\sqrt{3}/3$
(16) -1 (17) $\sqrt{3}/2$ (18) 1
(19) -1 (20) $-\sqrt{3}/2$ (21) undefined
(22) $-\sqrt{2}$ (23) $-\sqrt{2}$ (24) -2

Page 107
(1) 1 (2) $\sqrt{3}/2$ (3) undefined
(4) $1/2$ (5) $-\sqrt{3}$ (6) $\sqrt{3}$
(7) $-\sqrt{3}$ (8) -2 (9) 2
(10) $-1/2$ (11) 1 (12) $-\sqrt{2}$
(13) -1 (14) 1 (15) $-\sqrt{3}/3$
(16) -1 (17) $-\sqrt{2}$ (18) $-1/2$
(19) 0 (20) 0 (21) 0
(22) $-\sqrt{3}$ (23) $-1/2$ (24) $-\sqrt{3}$

Page 108
(1) $-\sqrt{2}$ (2) -1 (3) $-\sqrt{3}$
(4) -1 (5) undefined (6) 0
(7) -1 (8) $\sqrt{3}/2$ (9) $\sqrt{3}/3$
(10) $\sqrt{3}/3$ (11) $\sqrt{2}$ (12) $-1/2$
(13) undefined (14) $-\sqrt{3}$ (15) 1
(16) $-1/2$ (17) 1 (18) $-\sqrt{3}/3$
(19) $2\sqrt{3}/3$ (20) -1 (21) $\sqrt{2}$
(22) $1/2$ (23) 2 (24) $\sqrt{2}/2$

Chapter 8 Answers:

Page 111
(1) 120°, 240° (2) 150°, 330° (3) 0°, 180°
(4) 60°, 300° (5) 45°, 315° (6) 225°, 315°
(7) 210°, 330° (8) 225°, 315°
(9) 150°, 210° (10) 135°, 315°
(11) 120°, 240° (12) 120°, 240°
(13) 150°, 210° (14) 60°, 300°
(15) 210°, 330° (16) 45°, 315°
(17) 90°, 270° (18) 60°, 120° (19) 0°
(20) 0° (21) 150°, 330° (22) 135°, 315°
(23) 30°, 150° (24) 0°, 180°
Page 112
(1) 210°, 330° (2) 120°, 240°
(3) 120°, 240° (4) 90° (5) 120°, 240°
(6) 30°, 150° (7) 150°, 330° (8) 0°, 180°
(9) 240°, 300° (10) 0° (11) 60°, 120°
(12) 60°, 120° (13) 210°, 330°
(14) 150°, 330° (15) 45°, 315°
(16) 0°, 180° (17) 150°, 210°
(18) 240°, 300° (19) 45°, 135°
(20) 225°, 315° (21) 90°, 270°
(22) 0° (23) 30°, 210° (24) 60°, 300°
Page 113
(1) 60°, 120° (2) 60°, 300° (3) 45°, 225°
(4) 0°, 180° (5) 150°, 330° (6) 135°, 315°
(7) 45°, 225° (8) 135°, 225° (9) 135°, 225°
(10) 0°, 180° (11) 225°, 315°
(12) 120°, 300° (13) 150°, 330°
(14) 45°, 225° (15) 90°, 270°
(16) 210°, 330° (17) 90° (18) 45°, 315°
(19) 60°, 300° (20) 135°, 225°
(21) 240°, 300° (22) 30°, 210°
(23) 30°, 210° (24) 90°, 270°

Page 114
(1) 30°, 150° (2) 0°, 180° (3) 270°
(4) 135°, 225° (5) 150°, 330° (6) 90°
(7) 0°, 180° (8) 150°, 330° (9) 0°, 180°
(10) 0°, 180° (11) 0°, 180° (12) 120°, 240°
(13) 135°, 315° (14) 150°, 210°
(15) 45°, 135° (16) 60°, 300°
(17) 135°, 315° (18) 135°, 315°
(19) 60°, 120° (20) 60°, 240° (21) 0°
(22) 90°, 270° (23) 0°, 180° (24) 60°, 300°
Page 115
(1) 45°, 315° (2) 135°, 225° (3) 120°, 240°
(4) 30°, 150° (5) 0° (6) 90°, 270°
(7) 150°, 330° (8) 30°, 150° (9) 0°, 180°
(10) 270° (11) 0° (12) 45°, 135°
(13) 240°, 300° (14) 90°, 270°
(15) 210°, 330° (16) 45°, 135° (17) 0°
(18) 135°, 315° (19) 135°, 225°
(20) 120°, 240° (21) 30°, 330°
(22) 210°, 330° (23) 150°, 210°
(24) 60°, 120°
Page 116
(1) 135°, 315° (2) 30°, 210° (3) 30°, 210°
(4) 150°, 330° (5) 135°, 315° (6) 90°
(7) 150°, 330° (8) 60°, 240° (9) 30°, 150°
(10) 120°, 240° (11) 60°, 300°
(12) 45°, 135° (13) 60°, 120°
(14) 120°, 300° (15) 90°, 270°
(16) 90°, 270° (17) 60°, 120°
(18) 60°, 120° (19) 0°, 180°
(20) 120°, 240° (21) 90°, 270°
(22) 30°, 210° (23) 210°, 330°
(24) 135°, 225°

Page 117
(1) 120°, 300° (2) 240°, 300° (3) 270°
(4) 45°, 315° (5) 60°, 300° (6) 150°, 210°
(7) 0° (8) 150°, 210° (9) 120°, 300°
(10) 225°, 315° (11) 90°, 270°
(12) 150°, 330° (13) 30°, 210°
(14) 45°, 135° (15) 135°, 315°
(16) 45°, 315° (17) 45°, 315° (18) 0°, 180°
(19) 150°, 330° (20) 60°, 300°
(21) 240°, 300° (22) 45°, 225° (23) 270°
(24) 210°, 330°

Page 118
(1) 90° (2) 0°, 180° (3) 240°, 300°
(4) 45°, 225° (5) 150°, 330° (6) 135°, 315°
(7) 0° (8) 30°, 150° (9) 0°, 180°
(10) 135°, 315° (11) 90°, 270°
(12) 30°, 150° (13) 90°, 270°
(14) 120°, 240° (15) 135°, 225°
(16) 120°, 300° (17) 225°, 315° (18) 90°
(19) 120°, 240° (20) 0°, 180° (21) 0°, 180°
(22) 210°, 330° (23) 270° (24) 135°, 315°

Page 119
(1) 270° (2) 150°, 210° (3) 135°, 315°
(4) 0°, 180° (5) 0° (6) 0°, 180° (7) 270°
(8) 30°, 150° (9) 135°, 315°
(10) 0°, 180° (11) 210°, 330°
(12) 60°, 300° (13) 120°, 240°
(14) 60°, 300° (15) 270° (16) 45°, 135°
(17) 180° (18) 45°, 315° (19) 225°, 315°
(20) 90°, 270° (21) 135°, 315°
(22) 0°, 180° (23) 0° (24) 0°, 180°

Page 120
(1) 225°, 315° (2) 135°, 315° (3) 90°
(4) 90° (5) 0°, 180° (6) 135°, 315°
(7) 270° (8) 30°, 150° (9) 150°, 330°
(10) 210°, 330° (11) 120°, 240°
(12) 45°, 225° (13) 150°, 330°
(14) 225°, 315° (15) 240°, 300°
(16) 30°, 210° (17) 45°, 135° (18) 0°, 180°
(19) 150°, 330° (20) 90°, 270°
(21) 0°, 180° (22) 225°, 315°
(23) 30°, 150° (24) 240°, 300°

Page 121
(1) 90°, 270° (2) 30°, 210° (3) 30°, 330°
(4) 120°, 240° (5) 60°, 300° (6) 0°
(7) 45°, 225° (8) 120°, 240° (9) 45°, 315°
(10) 210°, 330° (11) 135°, 315°
(12) 150°, 210° (13) 0°, 180° (14) 0°
(15) 0°, 180° (16) 120°, 240°
(17) 135°, 315° (18) 45°, 225°
(19) 120°, 240° (20) 0°, 180°
(21) 60°, 120° (22) 90°, 270°
(23) 150°, 330° (24) 135°, 225°

Page 122
(1) 45°, 135° (2) 30°, 210° (3) 30°, 150°
(4) 0°, 180° (5) 30°, 150° (6) 210°, 330°
(7) 60°, 300° (8) 180° (9) 0°, 180°
(10) 60°, 300° (11) 30°, 150°
(12) 90°, 270° (13) 30°, 330°
(14) 60°, 300° (15) 0°, 180° (16) 0°, 180°
(17) 240°, 300° (18) 30°, 330°
(19) 45°, 315° (20) 45°, 315°
(21) 120°, 240° (22) 225°, 315° (23) 180°
(24) 30°, 150°

Chapter 9 Answers:

Page 130
(1) $2\sqrt{3}/3$ (2) 13 (3) 60° (4) 60°
Page 131
(1) $\sqrt{3}$ (2) 8 (3) 45° (4) 5
Page 132
(1) 45° (2) 45° (3) $\sqrt{5}$ (4) 3
Page 133
(1) 2 (2) $\sqrt{3}$ (3) 120° (4) $\sqrt{2}$
Page 134
(1) $\sqrt{2}$ (2) 105° (3) 60° (4) $\sqrt{3}$
Page 135
(1) 8 (2) $\sqrt{6}$ (3) 30° (4) $2\sqrt{6}$
Page 136
(1) 45° (2) $3\sqrt{3}$ (3) 135° (4) 120°

Page 137
(1) 4 (2) $\sqrt{2}$ (3) 45° (4) 14
Page 138
(1) $\sqrt{5}$ (2) 6 (3) 30° (4) $4\sqrt{2}$
Page 139
(1) $\sqrt{15}$ (2) 7 (3) 7 (4) $\sqrt{5}$
Page 140
(1) 8 (2) $\sqrt{6}$ (3) 120° (4) 120°
Page 141
(1) $7\sqrt{3}$ (2) 75° (3) $4\sqrt{2}$ (4) 135°

Chapter 10 Answers:

Page 150
(1) $\frac{\sqrt{6}+\sqrt{2}}{4}$ (2) $\sqrt{3}-2$ (3) $\sqrt{6}-\sqrt{2}$
(4) $\frac{\sqrt{2}-\sqrt{6}}{4}$ (5) $2+\sqrt{3}$
Page 151
(1) $-\sqrt{6}-\sqrt{2}$ (2) $\frac{-\sqrt{6}-\sqrt{2}}{4}$ (3) $\frac{\sqrt{6}+\sqrt{2}}{4}$
(4) $-2-\sqrt{3}$ (5) $\sqrt{2}-\sqrt{6}$
Page 152
(1) $\frac{\sqrt{6}+\sqrt{2}}{4}$ (2) $2+\sqrt{3}$ (3) $\frac{\sqrt{6}-\sqrt{2}}{4}$
(4) $\sqrt{2}-\sqrt{6}$ (5) $-2-\sqrt{3}$
Page 153
(1) $\frac{\sqrt{6}+\sqrt{2}}{4}$ (2) $-\sqrt{6}-\sqrt{2}$ (3) $-\sqrt{6}-\sqrt{2}$
(4) $2+\sqrt{3}$ (5) $-2-\sqrt{3}$
Page 154
(1) $\sqrt{3}$ (2) -1 (3) 2 (4) 0 (5) $\frac{2\sqrt{3}}{3}$
Page 155
(1) 0 (2) -2 (3) undefined
(4) undefined (5) 0
Page 156
(1) $\frac{\sqrt{2+\sqrt{2}}}{2}$ (2) $\frac{\sqrt{6}-\sqrt{2}}{4}$ (3) $1+\sqrt{2}$
(4) $2-\sqrt{3}$ (5) $-\sqrt{4+2\sqrt{2}}$

Page 157
(1) $\frac{1}{2}\sqrt{\frac{4-\sqrt{6}-\sqrt{2}}{2}}$ (2) $\frac{1}{2}\sqrt{\frac{4+\sqrt{6}-\sqrt{2}}{2}}$ (3) $1+\sqrt{2}$
(4) $\sqrt{4+2\sqrt{2}}$ (5) $2-\sqrt{3}$
Page 158
(1) $\pm\frac{4}{5}$ (2) $\pm\sqrt{5}$ (3) $\pm\frac{5}{13}$
(4) $\pm2\sqrt{2}$ (5) $\pm\frac{2\sqrt{2}}{3}$
Page 159
(1) $\pm\sqrt{3}$ (2) $\pm\frac{\sqrt{7}}{4}$ (3) $\pm\frac{\sqrt{5}}{2}$
(4) $\pm\frac{\sqrt{5}}{3}$ (5) $\pm\sqrt{3}$

Chapter 11 Answers:

Page 169
(1) 240°, 300° (2) 45°, 315° (3) 135°, 315°
(4) 30°, 330° (5) 105°, 165°
(6) 45°, 135°, 225°, 315°
Page 170
(1) 60°, 150° (2) 480°, 600° (3) 40°, 100°
(4) 15°, 75°, 105°, 165° (5) 0°, 270°
(6) 40°, 80°
Page 171
(1) 45°, 315° (2) 0°, 180° (3) 0°, 90°
(4) 75°, 105° (5) 112.5°, 157.5°
(6) 90°, 270°
Page 172
(1) 30°, 150°, 210°, 330°
(2) 30°, 60°, 120°, 150°
(3) 0°, 120°, 180°, 240° (4) 120°, 240°
(5) 30°, 90°, 150°, 270° (6) 225°, 315°
Page 173
(1) 90°, 210°, 330° (2) 30°, 330°
(3) 90°, 210°, 330° (4) 60°, 150°
Page 174
(1) 45°, 315° (2) 10°, 30°, 50°
(3) 0°, 300° (4) 30°, 150°

Made in the USA
Middletown, DE
11 January 2024

47660923R00104